Step by Step in Plant Maintenance and Works Services

AF654638

Charles Emmanuel Oppon,
Albert Kumi Arkoh,
Emmanuel Agbo-Tei,
Charles Emmanuel Kojo Oppon

BlueRose ONE.com
Stories Matter
NewDelhi • London

BLUEROSE PUBLISHERS
India | U.K.

BlueRose ONE.com
Stories Matter
NewDelhi • London

For permissions requests or inquiries regarding this publication,
please contact:

BLUEROSE PUBLISHERS
www.BlueRoseONE.com
info@bluerosepublishers.com
+91 8882 898 898
+4407342408967

ISBN: 978-93-7139-303-4

Cover Design: [Emmanuel Elikplim Akyea]
Typesetting: [Charles Emmanuel Oppon]

First Edition: January 2026

DEDICATION

This book is dedicated to our beloved families, whose unwavering support, patience, and encouragement made this journey possible.

In addition, to all engineers around the globe — the thinkers, builders, and problem-solvers — whose dedication continues to shape a better, sustainable world for generations to come.

PREFACE

Plant maintenance and works services are fundamental to ensuring the efficient and uninterrupted operation of industrial, commercial, and utility facilities. As technology evolves and operational demands increase, the importance of a structured approach to maintenance and the effective management of works services becomes even more critical. This book, "*Plant Maintenance and Works Services*", has been meticulously developed to serve as a comprehensive guide for students, engineers, technicians, and industry professionals.

The objective of this book is to provide a clear and detailed understanding of the principles, practices, and strategies involved in plant maintenance. It explores the technical and managerial aspects of maintaining various types of equipment, facilities, and systems, from routine inspections and preventative measures to advanced diagnostic and repair techniques. The book also delves into the critical field of works services, encompassing utilities such as heating, ventilation, air conditioning, electrical systems, plumbing, and more.

Key features of this book include:

- **Comprehensive Coverage**: Detailed discussions on topics ranging from maintenance strategies and equipment reliability to works services and safety protocols.
- **Practical Insights**: Real-world examples, troubleshooting techniques, and step-by-step procedures to bridge the gap between theory and application.
- **Modern Approaches**: Emphasis on predictive maintenance, condition monitoring, and sustainable practices to align with contemporary industry trends.
- **Illustrative Content**: Diagrams, flowcharts, and tables to simplify complex concepts and facilitate better understanding.
- **Applicability**: Case studies and examples from various industries, including manufacturing, energy, and utilities, to highlight the versatility of maintenance practices.

This book is intended to serve as a valuable resource for both learners and seasoned practitioners. For students, it offers foundational knowledge to prepare them for careers in plant maintenance and works management. For professionals, it provides insights into modern techniques and emerging trends to enhance their expertise and adapt to evolving challenges.

I extend my gratitude to the experts, practitioners, and educators whose insights and experiences have enriched the content of this book. It is my hope that this work serves as a trusted reference for all those committed to the art and science of plant maintenance and works services, contributing to safer, more efficient, and more sustainable industrial operations.

Engr. Prof. Charles Emmanuel Oppon
2025

FOREWORD

In today's fast-paced industrial landscape, the reliability and efficiency of plants and facilities are critical to sustaining productivity, minimizing downtime, and ensuring safety. Effective plant maintenance and the management of works services are no longer auxiliary functions but essential pillars that support the overall success of any organization. The publication of *Plant Maintenance and Works Services* is both timely and invaluable, as it addresses the pressing need for a comprehensive resource in this vital field.

This book provides a thorough exploration of maintenance strategies, techniques, and tools, emphasizing both the technical and managerial aspects of plant operations. It bridges the knowledge gap by covering a wide array of topics, including preventive and predictive maintenance, fault diagnosis, utilities management, and modern advancements like condition monitoring and automation. Each chapter has been thoughtfully crafted to ensure clarity, practicality, and relevance, making this book a treasure trove for professionals and learners alike.

What sets this book apart is its holistic approach to plant maintenance and works services. It not only discusses the mechanical and technical facets but also delves into the operational and sustainability challenges faced by contemporary industries. The integration of real-world examples, step-by-step procedures, and illustrative content enhances its utility as a practical guide.

As industries evolve to meet the demands of digitalization and sustainability, this book equips readers with the skills and knowledge to navigate these changes effectively. Whether you are an engineer, technician, manager, or student, the insights contained within these pages will serve as a foundation for achieving excellence in plant maintenance and works services.

I commend the authors for compiling such a comprehensive and insightful work. This book is not only a reflection of their expertise but also a testament to their commitment to advancing the field. It is a must-read for anyone dedicated to mastering the art and science of maintaining and managing plants and facilities.

Dr. Samuel Boahen
Kwame Nkrumah University of Science & Technology
Kumasi, Ghana
2025

ACKNOWLEDGEMENT

We give all glory and honor to Almighty God, whose wisdom, strength, and grace have guided us throughout the writing of this book. It is by His divine inspiration and unwavering presence that we were able to bring this work to completion.

Through every challenge, delay, and breakthrough, we felt His hand leading us, and for that, we are truly grateful. May this work serve its purpose and bring value to all who read it — to the glory of God.

TABLE OF CONTENTS

CHAPTER ONE

1.0 FUNDAMENTALS AND TERMINOLOGY OF MAINTENANCE SERVICE

1.1 INTRODUCTION

Plant Maintenance and Work Services refer to the activities and services involved in ensuring the efficient and continuous operation of industrial plants and facilities. These services are essential for preventing downtime, improving equipment performance, and ensuring safety and regulatory compliance.

1.2 WHAT IS PLANT MAINTENANCE?

Plant Maintenance involves a range of activities aimed at keeping machinery, equipment, and facilities in optimal working condition. It includes:

1.3 DUTIES OF MAINTENANCE DEPARTMENT

The maintenance department plays a crucial role in ensuring the smooth operation of an organization's facilities, equipment, and assets. Its primary duties include:

1.3.1 Preventive Maintenance

- Developing and implementing a preventive maintenance schedule for equipment and machinery.
- Conducting regular inspections, cleaning, and adjustments to minimize wear and tear.
- Ensuring equipment reliability by proactively addressing potential issues before they lead to breakdowns.

1.3.2 Corrective Maintenance

- Diagnosing and repairing equipment or machinery that breaks down unexpectedly.
- Ensuring quick response times to minimize downtime and restore operations as soon as possible.
- Managing emergency repairs and troubleshooting critical issues efficiently.

1.3.3 Facility Maintenance

- Overseeing the maintenance of buildings, grounds, and infrastructure (plumbing, electrical, HVAC systems, etc.).
- Ensuring that facilities comply with health, safety, and environmental regulations.
- Managing the upkeep of the physical environment, including landscaping, lighting, and signage.

1.3.4 Asset Management

- Tracking and managing the lifecycle of equipment, machinery, and other assets.
- Maintaining accurate records of maintenance history, repairs, and replacements.
- Making recommendations for equipment upgrades or replacements based on performance and cost-efficiency.

1.3.5 Inventory and Spare Parts Management

- Ensuring adequate stock levels of essential spare parts and maintenance supplies.
- Coordinating with suppliers to restock inventory and negotiate pricing for maintenance materials.
- Tracking inventory usage to optimize stock levels and reduce unnecessary expenses.

1.3.6 Safety Compliance and Risk Management

- Implementing safety procedures to protect workers and comply with safety regulations.
- Conducting regular safety audits and risk assessments.
- Training staff on safety protocols, proper handling of tools, and emergency response procedures.

1.3.7 Documentation and Record-Keeping

- Keeping detailed records of all maintenance activities, including repairs, inspections, and equipment usage.
- Documenting issues, resolutions, and preventive measures to streamline future maintenance work.
- Ensuring compliance with regulatory and internal documentation standards.

1.3.8 Cost Management and Budgeting

- Developing and managing the maintenance department's budget.
- Controlling costs by planning cost-effective maintenance solutions and minimizing waste.
- Analysing expenses and identifying areas to reduce costs while maintaining quality.

1.3.9 Vendor and Contractor Management

- Coordinating with third-party vendors or contractors for specialized maintenance tasks.
- Negotiating service agreements and monitoring performance to ensure high standards.
- Managing contracts and ensuring services are completed on schedule and within budget.

1.3.10 Support for Continuous Improvement Initiatives

- Implementing improvements to maintenance practices, procedures, and technology.
- Analysing maintenance data to identify trends and opportunities for process optimization.
- Supporting the organization's goals of efficiency and productivity by enhancing equipment reliability.

1.3.11 Staff Training and Development

- Training maintenance personnel on the latest maintenance techniques and best practices.
- Ensuring staff are knowledgeable about the operation and maintenance of new equipment.
- Supporting professional development through ongoing training, certifications, and performance evaluations.

1.3.12 Summary

The maintenance department's efforts ensure that facilities and equipment function efficiently, thereby reducing downtime, improving safety, and ultimately contributing to the organization's overall productivity and success.

1.4 PREVENTIVE MAINTENANCE

- Regularly scheduled inspections and servicing to prevent unexpected failures.
- Examples of preventive maintenance may include but not limited to lubrication, adjustments, cleaning, and parts replacement based on equipment usage or time intervals.

In conclusion, preventive maintenance tasks are proactive measures aimed at reducing equipment failures, extending the lifespan of assets, and avoiding unexpected downtime. These tasks are performed regularly to keep machinery, equipment, and systems in optimal working condition. Here are some common preventive maintenance tasks:

1.4.1 Inspections and Testing

- Regularly inspecting equipment, machinery, and systems for signs of wear, damage, or misalignment.
- Conducting routine diagnostic tests to detect potential issues before they become problems.
- Checking gauges, meters, and other indicators to ensure that equipment operates within safe limits.

1.4.2 Lubrication of Moving Parts

- Applying lubricants to moving parts such as: bearings, gears, chains, and joints to reduce friction and wear.
- Checking lubricant levels and refilling or replacing oils and greases as required.
- Cleaning and re-lubricating components to prevent contaminants from causing abrasion or wear.

1.4.3 Cleaning and Dusting

- Removing dust, dirt, and debris from equipment surfaces, internal components, and electrical parts.
- Cleaning air filters, fans, vents, and cooling systems to prevent overheating and improve efficiency.
- Ensuring that workspaces are free from clutter and contaminants that can affect machinery.

1.4.4 Tightening Bolts, Screws, and Fasteners

- Checking and tightening bolts, screws, and other fasteners that may loosen over time due to vibrations.
- Inspecting structural supports and reinforcements to prevent mechanical failures.
- Replacing worn or damaged fasteners as required.

1.4.5 Replacing Worn Parts

- Scheduling replacement of parts that wear out frequently, like belts, bearings, seals, and gaskets.
- Replacing components according to their expected lifespan, even if they have not failed.
- Keeping spare parts on hand to ensure replacements are readily available when needed.

1.4.6 Calibration and Adjustment

- Regularly calibrating sensors, gauges, and meters to ensure accurate readings.
- Adjusting settings or alignment of equipment to maintain operational accuracy.
- Checking and recalibrating equipment used for measurement or sensitive tasks to ensure precision.

1.4.7 Electrical System Checks

- Inspecting wiring, connectors, and electrical panels for signs of wear, corrosion, or damage.
- Testing battery backup systems, circuit breakers, and safety switches to ensure they function properly.
- Cleaning and tightening electrical connections to prevent shorts or power issues.

1.4.8 Fluid Checks and Refills

- Checking levels of essential fluids like oil, coolant, hydraulic fluid, and fuel.
- Refilling fluids to ensure that systems function smoothly and to prevent overheating or damage.
- Testing fluid quality to identify contamination or deterioration that may affect performance.

1.4.9 Updating Software and Firmware

- Installing updates for software or firmware on equipment with computerized controls.
- Running system checks to ensure compatibility and optimal functionality with the latest versions.
- Checking for cybersecurity vulnerabilities and updating protocols to maintain security.

1.4.10 Cleaning or Replacing Filters

- Inspecting and replacing air, oil, and fluid filters regularly to maintain clean operations.
- Cleaning reusable filters to remove dust, dirt, and contaminants.
- Ensuring proper ventilation and airflow by keeping filters clean and unblocked.

1.4.11 Testing Safety Equipment

- Checking safety devices like alarms, emergency stop buttons, fire extinguishers, and sprinklers.
- Inspecting personal protective equipment (PPE) such as gloves, helmets, and goggles.
- Conducting safety drills and simulations to ensure preparedness in emergencies.

1.4.12 Battery Inspection and Replacement

- Inspecting batteries in backup systems, portable devices, and critical equipment.
- Checking for corrosion, leaks, and ensuring proper charging and discharging.
- Replacing old or weak batteries to avoid power interruptions.

1.4.13 System Backups

- Backing up data in computerized systems to ensure important information is not lost.
- Creating redundancy plans to allow rapid recovery if systems fail.
- Testing backup procedures to ensure data recovery is possible in emergencies.

1.4.14 Condition Monitoring

- Using sensors or monitoring tools to track conditions like temperature, vibration, and pressure.
- Analysing data trends to predict when components might need attention or replacement.
- Adjusting maintenance schedules based on real-time monitoring insights.

1.4.15 Documentation and Record Keeping

- Keeping detailed records of maintenance activities, inspections, and replacements.
- Tracking maintenance schedules and recording completion to ensure nothing is missed.

- Using records to analyse patterns and improve preventive maintenance strategies.

1.4.16 Training and Skill Development

- Providing regular training for staff on how to identify early warning signs of potential issues.
- Updating staff on the latest maintenance procedures and safety protocols.
- Ensuring that team members know how to perform basic preventive maintenance tasks.

1.4.17 Summary

In summary, preventive maintenance helps to improve reliability, efficiency, and safety in industrial settings. By regularly performing these tasks, organizations can significantly reduce the risk of costly breakdowns and extend the lifespan of their equipment.

1.5 PREDICTIVE MAINTENANCE

- Using advanced techniques and tools to predict equipment failures before they happen.
- **Examples** – Vibration analysis, thermography, oil analysis, and ultrasonic testing.

Predictive maintenance (PdM) uses data analysis to identify operational anomalies and potential equipment defects, thereby, enabling timely repairs before failures occur. It aims to minimize maintenance frequency, avoiding unplanned outages and unnecessary.

Predictive maintenance tasks are based on monitoring the actual condition and performance of equipment to predict when maintenance is needed. This approach uses data from sensors, analytics, and advanced monitoring technologies to identify potential failures before they occur, reducing unplanned downtime and maintenance costs. Here are common predictive maintenance tasks:

1.5.1 Vibration Analysis

- Monitoring vibration levels in machinery to detect imbalances, misalignment, or bearing wear.
- Analysing vibration data trends to identify early signs of mechanical issues.
- Scheduling repairs or part replacements when abnormal vibration patterns are detected.

1.5.2 Thermography (Infrared Monitoring)

- Using infrared cameras to monitor heat signatures and detect hot spots in electrical systems, motors, and rotating equipment.

- Identifying areas of excessive heat that may indicate issues like friction, electrical resistance, or insulation breakdown.
- Scheduling maintenance when thermal anomalies are detected to prevent overheating and fires.

1.5.3 Ultrasonic Testing

- Using ultrasonic sensors to detect high-frequency sounds produced by equipment issues, like leaks, electrical discharge, or bearing defects.
- Pinpointing air or gas leaks, faulty bearings, and electrical arcing with ultrasound detection.
- Addressing problems immediately based on detected anomalies, preventing further equipment deterioration.

1.5.4 Oil Analysis

- Analysing the condition of lubricants in machinery to check for contaminants, viscosity changes, and wear particles.
- Tracking the presence of metal particles that may indicate wear and tear of internal components.
- Replacing or cleaning components based on lubricant quality data, extending equipment lifespan.

1.5.5 Electrical Insulation Testing

- Conducting insulation resistance testing on electrical systems to detect potential breakdowns or shorts.
- Monitoring insulation degradation over time, as this can lead to electrical failures.
- Replacing insulation or wiring as needed, based on test results.

1.5.6 Motor Current Analysis

- Monitoring electrical current in motors to detect issues like overloading, overheating, or winding damage.
- Tracking irregular current patterns that may indicate motor issues or potential failure.
- Scheduling motor maintenance when abnormal current levels are detected.

1.5.7 Condition Monitoring with IoT Sensors

- Installing Internet of Things (IoT) sensors on equipment to continuously track parameters such as temperature, humidity, pressure, and vibration.
- Analysing real-time data to identify changes that signal wear or performance degradation.
- Integrating IoT data with analytics to create predictive models and schedule maintenance proactively.

1.5.8 Wear Particle Analysis (Ferrography)

- Analysing particles in oil to determine the rate and type of wear occurring in machinery.
- Detecting abnormal wear patterns through the size, shape, and composition of particles.
- Replacing or repairing parts based on the condition of wear particles detected (Figure 1.1 (a) and (b)).

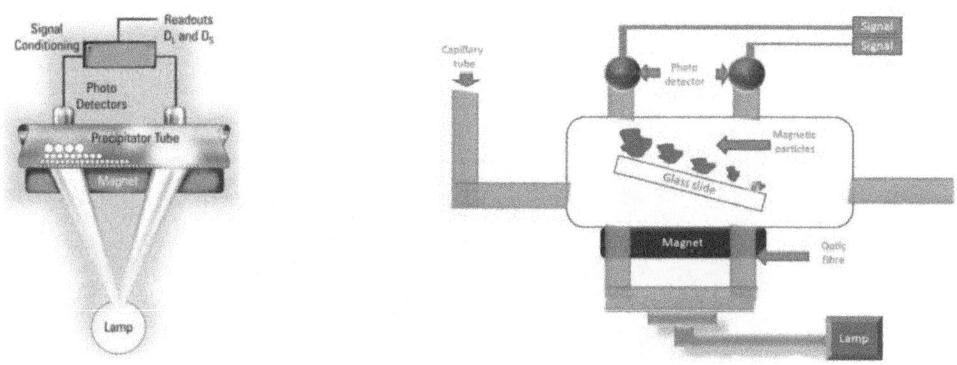

(a) (b)

Figure 1.1: Some Methods of Wear Particle Analysis

1.5.9 Acoustic Emission Testing

- Using acoustic sensors to detect stress waves emitted from cracks, leaks, or other structural issues in materials (Figure 1.2: Acoustic Emission Testing).
- Identifying structural issues in critical components like pressure vessels, pipelines, and turbines.
- Addressing potential failures before they lead to structural damage.

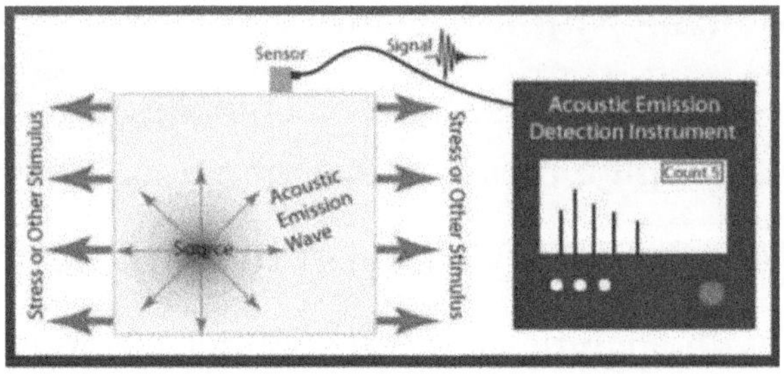

Figure 1.2: Acoustic Emission Testing

1.5.10 Pressure Monitoring

- Monitoring pressure levels in hydraulic and pneumatic systems to detect leaks, clogs, or pump issues.
- Detecting fluctuations that might indicate blockages, wear, or other issues affecting pressure.
- Replacing parts or scheduling system maintenance when abnormal pressure patterns are identified. The pressure monitoring system is as shown in Figure 1.3.

Figure 1.3: Pressure Monitoring System

1.5.11 Corrosion Monitoring

- Using corrosion sensors or probes to measure the rate of corrosion in pipes, tanks, and other metal equipment (Figure 1.4).
- Tracking corrosion trends over time to prevent structural failure due to material degradation.
- Replacing or treating materials as necessary based on corrosion levels.

Figure 1.4: Corrosion Monitoring Systems

1.5.12 Machine Learning and Predictive Analytics

- Applying machine learning algorithms to historical and real-time data to predict equipment failures.
- Using predictive models to analyse patterns in equipment usage, performance, and environmental factors.
- Automating maintenance schedules based on the output of predictive analytics.

1.5.13 Battery Health Monitoring

- Monitoring voltage, temperature, and current in battery systems to predict end-of-life.
- Identifying anomalies that may indicate impending failure or capacity degradation.
- Scheduling battery replacements or maintenance based on predictive data.

1.5.14 Visual Inspection with Drones or Robotics

- Using drones or robotic devices to visually inspect hard-to-reach areas for early signs of wear, corrosion, or structural damage.
- Detecting issues in equipment, structures, or pipelines in remote or hazardous locations.
- Scheduling maintenance activities based on visual inspection findings.

1.5.15 Water and Moisture Monitoring

- Monitoring moisture levels in electrical cabinets, buildings, or equipment to prevent corrosion, insulation breakdown, and mold.
- Tracking moisture patterns and water leaks to prevent damage to sensitive equipment.
- Repairing leaks or dehumidifying areas based on moisture data trends.

1.5.16 Alignment and Balancing Checks

- Monitoring alignment in rotating equipment, such as pumps and motors, to detect misalignment or imbalance.
- Using laser alignment tools to detect small shifts that could lead to increased wear or damage.
- Scheduling alignment corrections when minor misalignments are detected.

1.5.17 Asset Life Cycle Analysis

- Monitoring and analysing the entire life cycle of equipment to determine optimal replacement times.
- Using predictive models to calculate remaining useful life (RUL) for each asset.
- Replacing equipment before it reaches the end of its useful life, preventing unplanned downtime.

1.5.18 Air Quality Monitoring

- Using sensors to monitor air quality in facilities where sensitive machinery is located.
- Detecting particulate matter, dust, or pollutants that could degrade equipment performance.
- Scheduling maintenance or adjusting environmental controls based on air quality data.

1.5.19 Summary

In summary, predictive maintenance tasks allow industries to anticipate maintenance needs, optimize equipment performance, and minimize the risk of unplanned downtime. By monitoring real-time data and trends, organizations can take action at the right time, enhancing efficiency and reducing overall maintenance costs.

1.6 CORRECTIVE MAINTENANCE

- Repairs and restoration activities after a failure has occurred.
- Examples: Fixing broken components, adjusting misaligned parts, and replacing worn-out items.

Corrective maintenance tasks are activities performed to fix equipment, machinery, or systems after a fault or breakdown has occurred. These tasks are essential for restoring functionality and ensuring that operations can resume as quickly as possible. Here are some common corrective maintenance tasks:

1.6.1 Diagnosing Issues

- Troubleshooting equipment or system faults to identify the root cause.
- Running diagnostic tests to pinpoint malfunctions or inefficiencies.
- Using sensors, software, or manual inspection to assess the issue.

1.6.2 Repairing or Replacing Parts

- Replacing worn-out, damaged, or malfunctioning parts to restore equipment function.
- Repairing components such as belts, gears, motors, bearings, and seals.
- Fixing broken or damaged wiring, circuitry, or connectors in electrical systems.

1.6.3 Adjusting or Calibrating Equipment

- Realigning or recalibrating machinery to ensure optimal performance after repairs.
- Adjusting settings or controls that may have shifted or been incorrectly configured.
- Testing and recalibrating sensors, gauges, or meters for accuracy.

1.6.4 Cleaning and Lubricating Components

- Cleaning parts that may be causing a malfunction due to dirt, grime, or debris buildup.
- Applying lubrication to moving parts to prevent friction-related issues and ensure smooth operation.
- Removing rust or corrosion that may be affecting the equipment's function.

1.6.5 Software Troubleshooting and Updates

- Identifying and fixing software glitches or bugs in machinery with computerized systems.
- Updating firmware or software to ensure compatibility and enhance performance.
- Restoring settings or configurations that may have been disrupted by errors.

1.6.6 Replacing Consumables

- Replacing consumable items such as filters, gaskets, seals, and fuses that may have worn out.
- Refilling essential fluids like oil, coolant, or hydraulic fluids if levels are low.
- Checking and replacing batteries in equipment with electronic components.

1.6.7 Structural or Component Reinforcement

- Reinforcing parts or structures that have become weakened over time or due to damage.
- Welding, bolting, or fastening components that may have come loose or cracked.
- Adding supports or braces to stabilize machinery or prevent further damage.

1.6.8 Testing Equipment After Repairs

- Running test cycles to ensure that repairs were successful and that equipment functions properly.
- Monitoring performance to ensure the problem has been fully resolved and no secondary issues arise.
- Observing for abnormal sounds, vibrations, or other indicators of residual problems.

1.6.9 Documenting the Repair Process

- Recording the details of the fault, including the cause, actions taken, and any parts replaced.
- Updating maintenance logs to ensure that future maintenance activities are informed by past issues.
- Reporting recurring issues to determine if changes in preventive maintenance are necessary.

1.6.10 Coordinating with Vendors or Specialists

- Contacting specialized technicians if the repair requires expertise beyond the in-house team.
- Working with vendors to obtain specific parts or advice on handling complex repairs.
- Ensuring compliance with warranty or service agreements when using third-party contractors.

1.6.11 Inspecting Related Components

- Checking related or nearby components that may also be affected by the fault.
- Inspecting adjacent parts for wear, misalignment, or damage to prevent additional issues.
- Ensuring that interconnected systems are functioning correctly after the repair.

1.6.12 Summary

In summary, corrective maintenance tasks are often urgent and need to be completed promptly to minimize downtime. Effective corrective maintenance helps maintain productivity and prevent small issues from escalating into major problems.

1.7 ROUTINE MAINTENANCE

Routine maintenance refers to the regular, scheduled tasks performed to keep equipment, machinery, and systems in good working condition. These tasks are often simple but crucial for preventing unexpected breakdowns and ensuring the longevity and efficiency of assets. Routine Maintenance Tasks may include:

1.7.1 Equipment Inspections

Description – Regular visual inspections of machinery and equipment to check for signs of wear, damage, or malfunction.

Example: Inspecting belts, hoses, and filters in an HVAC system for wear and tear.

1.7.2 Lubrication

Description – Applying lubricants to moving parts to reduce friction and wear.

Example – Greasing bearings on a conveyor belt system to ensure smooth operation.

1.7.3 Filter Replacement

Description – Replacing filters in equipment to ensure efficient operation and prevent contamination.

Example – Changing the oil filter in an engine or the air filter in an HVAC system.

1.7.4 Tightening Bolts and Fasteners

Description – Ensuring that all bolts, screws, and fasteners are secure to prevent parts from coming loose during operation.

Example – Tightening the bolts on a machine's housing or casing to prevent vibration-related issues.

1.7.5 Cleaning

Description – Regular cleaning of equipment and components to remove dirt, dust, and debris that could cause malfunctions or reduce efficiency.

Example – Cleaning the fans and vents of a computer or industrial machine to prevent overheating.

1.7.6 Calibration

Description – Adjusting instruments and equipment to ensure accurate readings and performance.

Example – Calibrating a pressure gauge or a digital scale to ensure it provides accurate measurements.

1.7.7 Software Updates

Description – Updating the software and firmware of computerized systems to improve performance, fix bugs, and add new features.

Example – Installing updates on CNC machines or other automated systems to enhance functionality.

1.7.8 Battery Checks and Replacement

Description – Testing and replacing batteries in equipment to ensure they have sufficient charge and are operating correctly.

Example – Checking the batteries in emergency lighting systems and replacing them if necessary.

1.7.9 Testing Safety Systems

Description – Regularly testing safety features and systems to ensure they are functioning properly.

Example – Testing the emergency stop button on a production line or the fire alarm system in a building.

1.7.10 Fluid Level Checks

Description – Checking and replenishing fluid levels in machinery, such as oil, coolant, and hydraulic fluid.

Example – Checking the oil level in an engine or the coolant level in a cooling system and topping them up if needed.

1.7.11 Replacing Worn Parts

Description – Identifying and replacing parts that show signs of wear before they fail.

Example – Replacing worn-out fan belts or gaskets in a motor before they break down.

1.7.12 Electrical System Inspections

Description – Inspecting electrical systems and connections to prevent short circuits, overheating, or power failures.

Example – Checking wiring, circuit breakers, and fuses in a control panel.

1.7.13 Checking Tire Pressure and Condition

Description – Regularly checking the pressure and condition of tires on vehicles or machinery.

Example – Checking and adjusting the tire pressure on a forklift or checking for wear on the tires of a service vehicle.

1.7.14 Alignment Checks

Description – Ensuring that moving parts are properly aligned to prevent uneven wear and mechanical stress.

Example – Checking the alignment of conveyor belts or drive shafts in machinery.

1.7.15 Monitoring and Logging

Description – Recording the performance metrics of machinery and equipment during operation to identify trends and potential issues.

Example – Logging the temperature and vibration readings of a motor to monitor its condition over time.

1.7.16 Summary

These routine maintenance tasks are vital for ensuring that equipment operates efficiently, reducing the risk of unexpected failures, and extending the life of assets.

1.8 EMERGENCY MAINTENANCE

Immediate response to sudden equipment failures or breakdowns to restore operations.

Examples – On-the-spot repairs and replacements to minimize downtime.

Emergency Maintenance is an urgent, reactive form of maintenance performed in response to an unexpected failure or issue that poses an immediate threat to safety, productivity, or asset integrity.

This type of maintenance is critical for addressing situations where normal operations or safety protocols are compromised and immediate action is required to prevent further damage or to restore operational functionality.

1.8.1 Key Characteristics of Emergency Maintenance

Urgency – Emergency maintenance is performed immediately to mitigate risks, without delay for planning or scheduling.

Unplanned Nature – Unlike preventive or predictive maintenance, emergency maintenance arises unexpectedly and cannot be pre-scheduled.

Resource Intensive – Due to its sudden nature, emergency maintenance often requires reallocating resources, which may include personnel, tools, and equipment, to handle the issue promptly.

High Cost – Emergency maintenance is generally more expensive due to the premium on time and materials needed quickly and sometimes on an urgent basis.

Disruption to Operations – Emergency maintenance may cause partial or complete halting of operations, leading to potential downtime costs and loss of productivity.

1.8.2 Common Triggers for Emergency Maintenance

Emergency maintenance is generally initiated by events such as:

Critical Equipment Failure – Unexpected breakdowns of essential equipment that impact production or safety.

Safety Hazards – Situations like gas leaks, electrical hazards, or structural issues that pose immediate danger to personnel.

Environmental Conditions – Issues that may lead to environmental damage, like spills or emissions.

IT and Communication Failures – Critical failures in IT or communication systems that impact operational or safety protocols.

Example Scenarios

Manufacturing Plant – If a high-temperature furnace overheats and malfunctions, emergency maintenance would involve quickly assessing and resolving the problem to prevent equipment damage or fire.

Data Center – In the event of a cooling system failure that risks overheating server rooms, emergency maintenance would be needed to avoid data loss or hardware damage.

Facility Management – In a commercial building, if a fire alarm system malfunctions, emergency maintenance is required to ensure safety compliance and prevent potential hazards.

1.8.3 Steps in Emergency Maintenance

Emergency maintenance involves several critical steps:

- **Immediate Response** – Technicians are quickly mobilized to assess and diagnose the issue.
- **Safety Assessment** – Safety measures are put in place to protect personnel and assets before proceeding.
- **Quick Diagnosis and Repair** – Technicians rapidly identify the problem's cause and implement temporary or permanent repairs.
- **Documentation** – Once resolved, all actions taken, time spent, and resources used are documented for record-keeping and analysis.
- **Root Cause Analysis** – Following the event, teams may analyse the root cause to implement preventive measures and reduce recurrence.
- **Preventive Adjustments** – Post-incident, adjustments may be made to maintenance schedules or protocols to avoid similar emergencies.

1.8.4 Best Practices for Managing Emergency Maintenance

- **Preparedness** – Maintain a well-trained emergency response team with readily accessible tools and resources.

- **Clear Communication** – Establish protocols for fast, effective communication during emergencies to coordinate teams and inform stakeholders.
- **Resource Management** – Have contingency plans for quickly accessing critical spare parts and additional labor if needed.
- **Continuous Monitoring** – Use monitoring tools to detect early warning signs, which may help in reducing emergency incidents.
- **Documentation and Analysis** – Log all emergency maintenance activities and analyse data to develop strategies that reduce the frequency and impact of emergency incidents.

1.8.5 Balancing Emergency Maintenance with Other Maintenance Types

- While some emergencies are inevitable, excessive reliance on emergency maintenance can indicate deeper issues in asset management. To reduce the frequency of emergency events, organizations often:
- Increase preventive maintenance to identify and fix issues before they escalate.
- Implement predictive maintenance using condition-monitoring tools to address issues before failure.
- Improve training and protocols to better recognize and respond to warning signs.

1.8.6 Summary

In summary, emergency maintenance is a crucial response function for unexpected, high-priority incidents that threaten safety, productivity, or asset integrity. Effective management involves preparedness, clear communication, and documentation to minimize impact and improve future responses.

1.9 MAINTENANCE LOAD AND MAINTENANCE SYSTEM

"Maintenance load" and "maintenance system" are critical concepts in asset management, particularly in industrial settings. Below are explanations of each, along with how they relate to one another:

1.9.1 Maintenance Load

Maintenance load refers to the total volume and complexity of maintenance work required to keep all assets and equipment in optimal operating condition. It includes scheduled and unscheduled maintenance tasks, repairs, inspections, and replacements.

The maintenance load can be influenced by factors such as:

Equipment Age – Older equipment often requires more frequent and complex maintenance.

Operating Conditions – Harsh environments (e.g., extreme temperatures, high humidity) increase wear and tear.

Usage Intensity – Equipment used continuously or under heavy load typically has a higher maintenance load.

Type of Maintenance – Preventive, predictive, and corrective maintenance contribute differently to the overall load.

Asset Diversity – The complexity of maintaining different types of equipment with varied maintenance needs adds to the load.

Failures And Downtime – Unplanned breakdowns increase the maintenance load unexpectedly and urgently.

Managing maintenance load efficiently involves balancing the resources (labor, time, and budget) with the required tasks to maintain system reliability and performance.

1.9.2 Maintenance System

A maintenance system is a structured approach or framework for organizing, planning, and executing maintenance activities. It consists of tools, processes, software, and procedures designed to manage maintenance load effectively and ensure the longevity and reliability of assets. A well-designed maintenance system helps organizations streamline their maintenance operations, reduce costs, and improve equipment uptime.

1.9.3 Components of Maintenance System

1.9.3.1 Maintenance Planning and Scheduling

Maintenance planning and scheduling is a critical process in ensuring the efficiency, reliability, and longevity of equipment and facilities. It involves identifying maintenance needs, defining the scope of work, determining required resources such as tools, materials, and personnel, and allocating time for execution. Proper planning minimizes downtime, avoids unnecessary repairs, and helps prioritize tasks based on urgency and impact. Scheduling ensures that maintenance activities are carried out at optimal times, coordinating with production schedules to reduce disruptions. Together, planning and scheduling enhance overall operational performance, safety, and cost-effectiveness.

1.9.3.2 Maintenance Strategies

Preventive Maintenance (PM) – Scheduled, routine maintenance tasks based on time or usage.

Predictive Maintenance (PdM) – Maintenance based on real-time data and condition monitoring.

Corrective Maintenance (CM) – Reactive maintenance performed after an asset has failed or shown signs of failure.

Computerized Maintenance Management System (CMMS): Software that helps track, schedule, and document maintenance activities, inventory, and asset history.

Work Order Management – System for creating, assigning, and tracking maintenance tasks.

Resource Allocation – Planning for manpower, tools, spare parts, and budget based on the maintenance load.

Performance Tracking and KPIs – Monitoring performance metrics such as Mean Time Between Failures (MTBF) and Mean Time to Repair (MTTR) to evaluate the effectiveness of the maintenance system.

Safety and Compliance – Ensuring maintenance tasks are conducted safely and meet regulatory standards.

1.9.4 Relation Between Maintenance Load and Maintenance System

The maintenance system is designed to manage the maintenance load efficiently and effectively. The main goal is to balance resources with the workload, minimize unplanned downtime, and maximize equipment reliability.

Optimization – A well-designed maintenance system optimizes the maintenance load by ensuring that preventive and predictive maintenance are conducted effectively, reducing the occurrence of corrective maintenance.

Resource Management – Maintenance systems help allocate resources according to the maintenance load, enabling organizations to plan for manpower, budget, and spare parts based on anticipated needs.

Data-Driven Decisions – By using data from the CMMS, organizations can make informed decisions about asset management, helping to forecast maintenance load and adjust the system accordingly.

Prioritization – The maintenance system enables prioritization of tasks according to criticality, ensuring that high-priority equipment receives attention to prevent production disruptions.

Continuous Improvement – Maintenance systems facilitate continuous improvement by providing data on equipment performance, failure rates, and workload trends. This information can be used to adjust maintenance schedules, improve task execution, and balance the load over time.

1.9.5 Example of How a Maintenance System Manages Load

For instance, a manufacturing plant might have a maintenance system that uses CMMS software to manage maintenance load by:

Scheduling Preventive Maintenance – Automating regular inspections and tune-ups for equipment to avoid sudden failures.

Using Predictive Analytics – Tracking data such as vibration, temperature, and usage hours to predict when equipment might need servicing, thus smoothing out the maintenance load.

Managing Spare Parts Inventory – Ensuring that parts are available for scheduled maintenance without overstocking, reducing costs.

Allocating Labor – Assigning maintenance tasks to technicians based on workload forecasts, availability, and skill level.

Tracking Performance – Using KPIs to monitor maintenance efficiency and refine the approach if certain assets are taking up a disproportionately high maintenance load.

1.9.6 Summary

In summary, maintenance load is the demand on a maintenance team to keep equipment operational, while a maintenance system is the structure and tools used to manage this load effectively. A well-implemented maintenance system helps control and distribute the maintenance load, ensuring asset reliability and optimal resource use.

CHAPTER TWO

2.0 INTEGRATED MAINTENANCE MANAGEMENT: PLANNING, FEEDBACK, DOCUMENTATION, AND SYSTEM MECHANICS

2.1 THE PLANNING SYSTEM

The Planning System is an organized approach to defining, scheduling, and allocating resources for various activities within an organization. In an industrial or maintenance context, it's a structured framework that ensures that tasks are identified, prepared, prioritized, and completed efficiently, aligning with business objectives. A strong planning system is crucial for streamlining workflows, optimizing resources, minimizing costs, and ensuring reliability in operations. The key components of a planning system are as listed below:

2.1.1 Objective Definition

Establishing clear objectives and key performance indicators (KPIs) that guide what the planning system aims to achieve. Objectives may include increased production, improved equipment reliability, or enhanced safety.

2.1.2 Task Identification and Scope

- Determining specific tasks or activities required to achieve objectives. For maintenance, this includes preventive, predictive, corrective, and emergency maintenance tasks.
- Defining the scope of each task in terms of expected outcomes, needed resources, timelines, and responsible personnel.

2.1.3 Resource Allocation

- Identifying and assigning required resources (labor, tools, materials) to tasks to ensure work is completed without delays. This includes: personnel, specialized equipment, spare parts, and budget allocation.
- Aligning with personnel schedules and availability, skill requirements, and budget constraints.

2.1.4 Scheduling

- Developing a timeline that sequences tasks in an efficient order. This involves determining when tasks should occur to avoid clashes with other critical operations.
- Optimizing schedules to minimize disruptions, plan around peak operation times, and consider equipment downtime.

2.1.5 Priority Setting

- Prioritizing tasks based on criticality, safety risks, production impact, and compliance requirements.
- Using priority levels to allocate resources effectively, ensuring that high-priority issues are addressed first.

2.1.6 Workflow Optimization

- Designing workflows that reduce redundant steps, optimize movement, and minimize waste.
- Implementing processes and procedures that ensure that tasks flow smoothly from planning to completion, often using software to manage this process.

2.1.7 Documentation and Record-Keeping

- Maintaining records of tasks, resources used, schedules, and outcomes. This provides a database for analysis and helps track equipment history, compliance, and maintenance needs.
- Creating a foundation for predictive analysis, identifying recurring issues, and improving long-term planning.

2.1.8 Monitoring and Feedback

- Tracking performance against the initial plan, adjusting schedules and resources as needed.
- Gathering feedback post-task completion to assess plan effectiveness, worker efficiency, and areas for improvement.

2.1.9 Continuous Improvement

- Using performance data and feedback to improve future planning, resource allocation, and workflows. Adopting best practices and adapting based on trends and outcomes.
- Implementing new strategies, technologies, or methodologies based on past performance and changing operational demands.

2.1.10 Types of Planning in a Planning System

- **Strategic Planning** – Long-term, high-level planning aligned with organizational goals and objectives, focusing on achieving overall business success.
- **Tactical Planning** – Shorter-term planning focusing on implementing strategies through resource allocation and scheduling.
- **Operational Planning** – Day-to-day planning that includes scheduling specific tasks, assigning resources, and ensuring operational readiness.
- **Contingency Planning** – Preparing for unexpected situations, including equipment failures or supply chain issues, to ensure a quick response with minimal disruption.

2.1.11 Tools and Technology in Planning Systems

- **Enterprise Resource Planning (ERP)** – Integrates planning with overall business processes, including inventory management, scheduling, and finance.
- **Computerized Maintenance Management System (CMMS)** – Tracks maintenance tasks, schedules, inventory, and worker availability to optimize maintenance activities.
- **Project Management Software** – Manages complex projects, schedules, and team collaboration, providing real-time updates on task status and resource use.
- **Data Analytics and Predictive Tools** – Uses historical and real-time data to predict maintenance needs, optimize schedules, and minimize equipment failures.

2.1.12 Benefits of a Robust Planning System

- **Increased Efficiency** – Clear task allocation and optimized schedules reduce downtime and maximize productive time.
- **Resource Optimization** – Efficiently uses resources, minimizing waste and reducing operating costs.
- **Enhanced Safety** – A structured approach reduces the likelihood of hazards, ensuring compliance with safety regulations.

- **Improved Reliability** – Ensures that assets are well-maintained, enhancing reliability and reducing the likelihood of emergency maintenance.
- **Cost Savings** – Prevents unplanned costs through better preparation and proactive resource allocation.

2.1.13 Challenges in Implementing a Planning System

- **Complexity** – Large organizations may struggle to coordinate across departments, particularly in complex projects.
- **Resource Constraints** – Limited resources can make scheduling and prioritizing tasks challenging.
- **Resistance to Change** – Employees may resist adopting new planning tools or processes, requiring effective change management.
- **Unforeseen Events** – Unexpected disruptions can strain the planning system, necessitating flexibility and adaptability in plans.

2.1.14 Summary

A robust planning system integrates all aspects of an organization's operations, from maintenance to project management, creating a proactive approach that benefits productivity, efficiency, and cost-effectiveness.

1.2 USE OF FEEDBACK AND RECORDS IN MAINTENANCE MANAGEMENT

The use of feedback and records is a critical part of maintenance management as it enables continuous improvement, optimizes performance, and supports decision-making processes. By systematically capturing, analysing, and applying feedback and records, maintenance teams can effectively enhance the reliability, efficiency, and safety of equipment.

2.2.1 Capturing and Analysing Feedback

Feedback is the information provided by technicians, operators, or inspectors about the performance, issues, and operational status of equipment. It can be collected after every maintenance task or during routine inspections. Feedback provides real-time insights into equipment behavior, the effectiveness of maintenance procedures, and potential improvement areas.

Examples

- ***Technician Feedback*** – If technicians frequently report that a specific pump often fails despite routine maintenance, it may indicate underlying issues such as design flaws or inappropriate maintenance techniques. Analysing this feedback allows the maintenance team to adjust maintenance schedules, apply root cause analysis, or consider equipment redesign.
- ***Operator Feedback*** – Machine operators may report unusual noises, vibrations, or other early signs of potential breakdown. This feedback can prompt predictive maintenance measures, preventing unexpected failures.

2.2.2 Record-Keeping for Maintenance History

Maintenance records contain comprehensive information about each asset, including details on previous repairs, inspections, and performance issues. These records serve as an equipment history that aids in tracking reliability, spotting recurring failures, and planning long-term strategies.

Example

Maintenance History Records – A manufacturing plant tracks the repair history of a conveyor system and finds that the motor has failed three times within a year. The records indicate that each failure occurred after approximately three months of operation. This pattern helps the maintenance team recognize that the motor needs more frequent maintenance or possibly a replacement with a higher-capacity motor to handle the load demands.

2.2.3 Optimizing Maintenance Scheduling and Resources

Feedback and records inform maintenance scheduling by providing data on equipment condition and repair frequency. This information helps determine optimal intervals for preventive maintenance, predict resource needs, and reduce unnecessary maintenance on reliable equipment.

Example

Preventive Maintenance Scheduling – Records indicate that an HVAC system performs optimally with cleaning and filter replacement every four months rather than the previous six-month interval. By adjusting the schedule based on feedback, the maintenance team can maintain efficiency and prevent frequent breakdowns.

2.2.4 Enhancing Compliance and Safety Standards

Detailed records also assist with regulatory compliance and safety audits. Records verify that maintenance tasks meet safety standards, and feedback from technicians can highlight safety hazards that need immediate attention.

Example

Compliance and Safety: In industries with strict safety regulations, such as oil and gas, records verify compliance with inspection and maintenance intervals required by regulatory bodies. If feedback from an inspector highlights safety hazards like leaking pipes, records help document corrective actions taken, protecting the company from potential liabilities.

2.2.5 Root Cause Analysis and Continuous Improvement

Feedback and records play an essential role in root cause analysis by providing a data-backed approach to troubleshooting recurrent problems. Analysing patterns within records and applying feedback can lead to improvements in maintenance practices and equipment design.

Example

Root Cause Analysis: Suppose records indicate a recurring issue with a production line's hydraulic system. After analysing technician feedback and records of previous repairs, the maintenance team discovers that inadequate lubrication practices were causing frequent malfunctions. This insight prompts a revision of lubrication schedules and training for staff, leading to better performance and fewer breakdowns.

2.2.6 Summary

In conclusion, feedback and record-keeping are valuable for guiding data-driven decisions in maintenance. Properly maintained records and continuous feedback loops facilitate planning, improve asset reliability, enhance safety, and help the maintenance department optimize resources and costs.

2.3 INSTRUCTION AND OPERATING MANUAL IN MAINTENANCE MANAGEMENT

An instruction and operating manual is a crucial document provided by manufacturers, offering comprehensive guidance on how to install, operate, maintain, and troubleshoot equipment. These manuals serve as a primary reference for equipment operation and upkeep, helping ensure that machinery is used correctly, safely, and efficiently.

2.3.1 Key Elements of an Instruction and Operating Manual

- **Operating Procedures** – Detailed steps to operate the equipment safely, including startup, shutdown, and emergency procedures.
- **Maintenance Guidelines** – Recommendations for routine and preventive maintenance tasks, often including checklists for daily, weekly, monthly, and annual inspections.
- **Troubleshooting Tips** – Instructions on identifying and resolving common problems, often with troubleshooting flowcharts for systematic fault finding.
- **Safety Warnings and Precautions** – Information on potential hazards, including the correct use of protective gear and safety protocols.
- **Technical Specifications** – Information on parts, tolerances, operating ranges, and other technical parameters to ensure optimal operation.
- **Illustrations and Diagrams** – Visuals that support understanding, such as part diagrams, circuit layouts, or schematic drawings.

2.3.2 Importance of Instruction and Operating Manuals

1 Ensuring Correct Operation

The manual guides operators on the correct procedures to start, operate, and shut down equipment, which helps avoid improper handling that could lead to inefficiencies or breakdowns. For instance, an instruction manual for a CNC (Computer Numerical Control) machine would include step-by-step instructions on setting up the equipment, running a program, and powering down. Following these steps precisely prevents errors that could damage the machine or affect production quality.

2 Improving Safety and Reducing Risks

Operating manuals contain essential safety warnings and guidelines that protect operators from harm. This includes advice on using personal protective equipment (PPE) and warnings about hazardous materials or processes. For example, the manual for an industrial boiler may include safety steps to avoid overheating and pressure build-up, reducing the risk of accidents.

3 Guiding Routine Maintenance and Inspections

Routine maintenance sections within an operating manual provide schedules and procedures for routine tasks, helping the maintenance team keep the equipment in good working order. For example, a manual for a hydraulic press might specify daily inspections of hydraulic fluid levels, monthly checks on pressure relief valves, and annual servicing. These guidelines help ensure consistent, manufacturer-recommended care, prolonging equipment life and enhancing performance.

4 Supporting Troubleshooting and Problem Resolution

When equipment malfunctions, the troubleshooting section of the manual provides solutions to common issues, such as error codes or unusual sounds. This reduces downtime by enabling operators and technicians to quickly diagnose and resolve minor problems without waiting for specialized help. For instance, an air compressor manual might list error codes that indicate specific issues, such as low oil levels or clogged filters, along with solutions to address them.

5 Standardizing Training for New Operators

Instruction and operating manuals are valuable resources for training new staff on proper operation and maintenance procedures. Using the manual, trainers can teach consistent practices that align with the manufacturer's specifications, reducing variability in equipment handling. For example, new operators learning to use a 3D printer in a production setting can refer to the manual to understand setup, calibration, and printing procedures, ensuring consistency and quality.

6 Improving Inventory Management

Manuals often include a detailed list of parts and specifications, making it easier for the maintenance team to order the correct replacement parts, reducing equipment downtime. For example, a manual for a conveyor system may include the part numbers and specifications for belts, motors, and rollers, allowing the purchasing team to keep essential items in stock.

2.3.3 Examples of Instruction and Operating Manuals in Action

- **HVAC Systems** – The operating manual for an HVAC system may include instructions on thermostat settings, filter replacement, duct inspection, and refrigerant levels. These ensure the system functions efficiently, maintaining comfortable temperatures and minimizing energy consumption.
- **Forklift** – A forklift manual might specify procedures for safe operation, battery maintenance, and checking hydraulic systems. Following these instructions reduces the risk of breakdowns and ensures safe handling practices are followed on site.
- **Industrial Mixers** – For food processing mixers, the manual will provide instructions for disassembly and cleaning after use, safety precautions, and preventive maintenance such as lubricating moving parts. This promotes food safety and prevents wear from impacting machine performance.

2.3.4 Summary

In conclusion, instruction and operating manuals are foundational tools that support the correct, safe, and efficient use of equipment. They are central to training, maintenance planning, troubleshooting, and maintaining compliance with safety and quality standards, making them invaluable in any maintenance management system.

2.4 THE BASIC MECHANICS OF PLANNED MAINTENANCE SYSTEMS

A planned maintenance system is a structured approach to scheduling and performing maintenance tasks on equipment and facilities at predetermined intervals to prevent unexpected breakdowns and ensure optimal operation. Unlike reactive maintenance, which addresses issues as they arise, planned maintenance is proactive, aiming to extend equipment life, reduce downtime, and improve safety.

2.4.1 Key Components of a Planned Maintenance System

- **Scheduled Intervals** – Tasks are carried out on a regular schedule based on time (e.g., daily, weekly, monthly) or usage metrics (e.g., machine hours, production cycles). This helps prevent equipment from reaching a failure point.
- **Task Lists and Checklists** – Standardized procedures for each maintenance task ensure consistency and completeness. Checklists guide technicians on what to inspect, lubricate, replace, or repair.
- **Resource Allocation** – Planning for labor, tools, spare parts, and other resources required for maintenance ensures that these elements are available when needed.
- **Maintenance Records and Tracking** – Recording each maintenance action, including observations and repairs made, helps track the condition of the equipment and identifies recurring issues that may require deeper analysis.
- **Feedback Mechanism** – Input from technicians and operators about equipment performance post-maintenance allows adjustments to the maintenance plan as needed, ensuring it remains effective.
- **Work Order System** – A work order system organizes maintenance tasks, tracks progress, and serves as documentation for the actions taken on each asset.

2.4.2 Examples of Planned Maintenance Systems in Action

HVAC Systems in Commercial Buildings

- **Schedule** – HVAC filters are checked and replaced every quarter, while the entire system is inspected twice a year.
- **Task Lists** – During each check, technicians follow a checklist to inspect airflow, thermostat settings, refrigerant levels, and electrical connections.
- **Records** – The maintenance team logs findings from each inspection, noting any refrigerant leaks, worn parts, or airflow restrictions. This history helps forecast when major overhauls may be necessary.

2.4.3 Manufacturing Machinery in Production Lines

- **Schedule** – Preventive maintenance is scheduled monthly and based on production cycles. For instance, machines that run continuously may require more frequent lubrication.
- **Resource Allocation** – Parts such as bearings, belts, or filters are kept in stock, based on the frequency of replacements tracked in the maintenance history.
- **Work Orders and Tracking** – Each machine has a digital work order system for technicians to document maintenance tasks completed and machine status.

2.4.4 Vehicle Fleet in Logistics Companies

- **Schedule** – Vehicles undergo routine maintenance every 5,000 kilometers or quarterly, whichever comes first.
- **Checklist** – Inspections cover oil and filter changes, tire pressure, brake system checks, and fluid levels.
- **Tracking** – Data from each service is recorded, and patterns can reveal which vehicles may need earlier maintenance, such as brake replacements for those with more frequent urban driving.

2.4.5 Benefits of a Planned Maintenance System

- **Increased Equipment Lifespan** – By following a consistent maintenance schedule, equipment operates within optimal conditions, reducing wear and tear.
- **Reduced Downtime** – Scheduled maintenance helps identify and resolve issues before they lead to a failure, minimizing unexpected stoppages.

- **Enhanced Safety and Compliance** – Consistent maintenance addresses potential safety hazards, ensuring compliance with regulations and reducing the risk of accidents.

2.4.6 Feedback and Continuous Improvement

Feedback is essential to refining a planned maintenance system. Technicians can note recurring problems, such as frequent belt replacements in a conveyor system, and report these findings for further analysis. Based on this feedback, the maintenance team might adjust the inspection frequency, modify the task checklist, or recommend equipment upgrades.

2.4.7 Summary

A planned maintenance system integrates scheduling, task lists, resource planning, and feedback loops to maintain equipment reliability, safety, and efficiency. By systematically managing each of these elements, organizations can predict and prevent issues, reducing downtime and maintaining steady operations.

CHAPTER THREE

3.0 REGULATORY AND OPERATIONAL FRAMEWORKS IN MAINTENANCE MANAGEMENT

3.1 FACILITY REGISTERS

A facility register is a comprehensive record that catalogues all assets within a facility, including their location, specifications, maintenance history, and operational status. These registers serve as a vital reference for facility managers, technicians, and maintenance teams, offering a detailed overview of every piece of equipment and infrastructure within an organization. This organized information system enables better planning, faster troubleshooting, and efficient resource management.

3.1.1 Components of Facility Registers

- **Asset Identification** – Each asset is assigned a unique identification number or code, making it easy to locate specific equipment in the database.
- **Asset Details** – Essential details like model number, serial number, date of acquisition, specifications, manufacturer details, and location within the facility.
- **Maintenance History** – All records of past maintenance, repairs, inspections, and any adjustments made to the asset. This history helps in understanding recurring issues and planning preventive maintenance.
- **Operational Status** – Information on whether the asset is operational, under repair, or out of service, allowing facility managers to make informed decisions.
- **Replacement or Upgrade Schedule** – Expected dates for upgrades or replacements of the asset based on its lifecycle and current condition.

3.1.2 Examples of Facility Registers in Action

1 Healthcare Facility Register

- **Overview** – Hospitals and healthcare centers keep detailed registers of medical equipment, such as MRI machines, ultrasound devices, defibrillators, and patient beds.
- **Usage** – A facility register helps track each equipment's maintenance history and alert staff for necessary servicing or inspections. If a particular ultrasound machine has a

recurring calibration issue, it can be identified promptly and serviced to ensure patient safety.

- **Replacement Planning** – Older equipment nearing its end of life can be scheduled for replacement in advance, reducing the risk of unexpected downtime.

2 Educational Institution Facility Register

- **Overview** – Schools and universities may catalog assets like projectors, laboratory equipment, computers, desks, and HVAC systems.
- ***Usage*** – If a projector in a lecture hall malfunctions, the maintenance team can quickly look up its maintenance history to identify past issues. The register might reveal that the projector bulb is overdue for replacement, allowing quick corrective action.
- ***Planning for Upgrades*** – Educational facilities can use the register to keep track of computers or lab equipment that may need upgrades, ensuring the institution remains up-to-date with modern learning tools.

3 Manufacturing Plant Facility Register

- ***Overview*** – In manufacturing plants, facility registers include heavy machinery, conveyor belts, forklifts, generators, and HVAC systems.
- ***Usage*** – If a conveyor belt requires frequent repairs, the register will highlight its history of breakdowns, helping management decide if a replacement is more cost-effective.
- ***Preventive Maintenance*** – The facility register can notify maintenance teams of when machines are due for preventive maintenance tasks, reducing the chances of costly breakdowns during production.

4 Office Building Facility Register

- ***Overview*** – Office buildings maintain registers for elevators, security systems, fire alarms, lighting, and HVAC units.
- ***Usage*** – If an elevator malfunctions, the register helps facility management quickly access its service history, speeding up repairs.
- ***Compliance and Safety*** – Registers also track inspections and certifications required for compliance with safety regulations, such as fire alarm checks and electrical inspections.

3.1.3 Benefits of Facility Registers

- Informed Decision-Making – Facility registers provide a single source of truth, helping facility managers understand asset conditions, anticipate replacements, and allocate resources.
- Efficiency – By providing ready access to asset details, maintenance teams can perform faster, more accurate repairs and track down problems.
- Cost Savings – Facility registers enable better budgeting and resource allocation for preventive maintenance, reducing unexpected breakdowns and optimizing asset utilization.

3.1.4 Using Facility Registers in Modern Facilities

Facility registers are often implemented through Computerized Maintenance Management Systems (CMMS) or asset management software, making it easy to organize, update, and retrieve asset data across locations. By using these digital systems, facility managers can analyse trends, receive alerts for upcoming maintenance, and monitor the entire lifecycle of their assets.

3.1.5 Conclusion

Facility registers are critical for efficient facility management, ensuring all assets are accounted for, well-maintained, and prepared for both daily use and long-term operation. By centralizing information about each asset's details, condition, and maintenance requirements, registers support smooth facility operation, cost-effective maintenance, and effective resource planning.

3.2 INVENTORY IN MAINTENANCE MANAGEMENT

In maintenance management, inventory refers to the stock of materials, spare parts, tools, and equipment kept on hand to support the continuous and efficient operation of facilities. Maintaining an accurate and well-organized inventory helps prevent costly downtime, ensures timely repairs, and supports preventive and corrective maintenance tasks. Effective inventory management optimizes resource use, reduces waste, and helps in budgeting and planning maintenance operations.

3.2.1 Key Components of Maintenance Inventory

- **Spare Parts** – These include replacement parts for machinery and equipment, such as motors, belts, filters, and bearings, essential for unplanned repairs and preventive maintenance.
- **Consumables** – Items like lubricants, cleaning agents, welding rods, and other frequently used materials that are consumed during routine maintenance tasks.
- **Tools and Equipment** – Maintenance tools, diagnostic instruments, and specialized equipment required by maintenance personnel for repairs, inspections, and servicing.
- **Safety Stock** – Extra units of critical items kept on hand to cover unexpected needs and avoid delays due to supply chain disruptions.

3.2.2 Examples of Inventory in Different Industries

3.2.2.1 Manufacturing Plant Inventory

- **Overview** – Manufacturing plants maintain extensive inventories of spare parts for their machinery, including items like conveyor belts, motors, and hydraulic parts.
- **Usage** – If a conveyor belt malfunctions, the maintenance team can quickly replace it with a belt from the inventory, avoiding delays in production. By having these spare parts available, they minimize production downtime and keep the plant operating efficiently.
- **Safety Stock** – A manufacturing facility often keeps a minimum safety stock of essential parts like machine bearings and motors, reducing the risk of halted production if a part is unavailable.

3.2.2.2 Healthcare Facility Inventory

- **Overview** – Hospitals and healthcare centers rely on inventories for critical medical equipment parts and consumables, such as batteries for life-support machines, filters for air purifiers, and calibration equipment for diagnostic tools.
- **Usage** – For instance, if an MRI machine's coolant system requires maintenance, having the coolant and necessary tools in inventory allows for quick repairs without service interruptions.
- **Consumables** – Medical facilities keep consumables like disinfectants, gloves, and masks as part of their inventory for routine maintenance and cleaning of equipment, ensuring compliance with health standards.

3.2.2.3 Office Building Maintenance Inventory

- **Overview** – Large office buildings maintain an inventory of items necessary for HVAC systems, lighting, and plumbing maintenance, such as lightbulbs, filters, and plumbing seals.
- **Usage** – If a light fixture goes out in a common area, maintenance staff can quickly replace it using bulbs from inventory. Similarly, filters for HVAC systems are stocked for regular replacements, ensuring optimal air quality.
- **Preventive Maintenance** – Building managers use inventory items for routine maintenance tasks, such as replacing HVAC filters every few months to maintain energy efficiency and air quality.

3.2.2.4 Utility Company Inventory

- **Overview** – Utility companies maintain extensive inventories for their infrastructure needs, like power transformers, cables, poles, and various electrical components.
- **Usage** – During storm seasons, utility companies often rely on safety stock to quickly replace damaged transformers and lines, minimizing service interruptions.
- **Strategic Stock** – Utility companies may store key items in regional warehouses for rapid deployment, allowing them to address service disruptions and emergencies promptly.

3.2.3 Benefits of Effective Inventory Management

- **Reduced Downtime** – Having the right parts on hand reduces equipment downtime, as maintenance teams can perform immediate repairs rather than waiting for parts to be ordered and delivered.
- **Cost Savings** – By optimizing inventory levels, organizations avoid overstocking or understocking, saving on storage costs while ensuring that essential items are available.
- **Enhanced Maintenance Efficiency** – Maintenance teams can work more efficiently, knowing they have access to the tools and parts they need, reducing delays in repair or preventive maintenance tasks.
- **Better Budget Control** – Inventory management allows facility managers to budget for parts and supplies accurately, making it easier to allocate funds for future needs and reduce waste.

3.2.4 Inventory Management Techniques in Maintenance

- **Just-In-Time (JIT) Inventory** – JIT keeps inventory levels low, ordering items only when they are needed for upcoming maintenance tasks. This method reduces storage costs but requires reliable suppliers to avoid shortages.
- **ABC Analysis** – This technique categorizes inventory based on value and usage frequency, allowing managers to focus on maintaining high levels of essential items while minimizing low-priority stock.
- **Computerized Maintenance Management Systems (CMMS)** – Many organizations use CMMS software to track inventory levels, automate reordering, and manage stock usage. A CMMS can trigger reorders when inventory reaches predefined minimum levels.

3.2.5 Summary

Inventory management is an essential component of maintenance management, providing organizations with the necessary tools, spare parts, and consumables to ensure equipment reliability and operational efficiency. By carefully monitoring and managing inventory, maintenance departments can reduce downtime, improve maintenance response times, and optimize resource use, ultimately contributing to the facility's overall performance.

3.3 FACTORY LEGISLATION IN MAINTENANCE MANAGEMENT

Factory legislation refers to a set of laws and regulations enacted to ensure safe, fair, and healthy working conditions in industrial settings. These laws establish minimum standards for workplace safety, equipment maintenance, employee health, and environmental protection, helping reduce risks, improve operational efficiency, and protect workers' rights. Factory legislation often mandates that factories maintain specific standards, conduct regular inspections, and provide appropriate training, with non-compliance typically resulting in penalties or fines.

3.3.1 Key Areas of Factory Legislation

1. ***Occupational*** **Health and Safety** – Laws aimed at ensuring that employees have a safe work environment free from hazards. This includes rules on machine maintenance, use of protective equipment, and emergency preparedness.
2. **Environmental Regulations** – Standards related to waste disposal, emissions, and pollution control are essential for minimizing the environmental impact of factory operations.

3 **Fire Safety** – Regulations on fire prevention measures, emergency exits, fire alarms, and extinguishing equipment are critical for minimizing fire-related risks in factories.

4 **Machine Maintenance and Inspection** – Specific rules requiring routine inspection and maintenance of machinery and equipment to ensure they function safely and efficiently.

5 **Worker Rights and Welfare** – Laws ensuring fair wages, reasonable working hours, access to restrooms, and protections against discrimination or exploitation.

6 **Permit-to-Work System** – Many countries require permits for certain maintenance activities, such as working in confined spaces, handling hazardous chemicals, or conducting high-risk tasks.

3.3.2 Examples of Factory Legislation in Different Contexts

- ***Occupational Safety and Health Administration (OSHA) Standards*** – United States
- ***Overview*** – OSHA requires that factories in the U.S. follow specific safety standards to protect workers from potential hazards. For example, OSHA mandates regular inspections and maintenance of heavy machinery to prevent malfunctions that could cause harm to workers.
- ***Example*** – In a steel manufacturing plant, all press machines must be inspected monthly and maintained according to OSHA regulations. Failure to maintain or inspect the machines can lead to severe penalties for the company and increased accident risks.

3.3.3 Factories Act 1948 – India

- **Overview** – The Factories Act of 1948 governs worker health, safety, and welfare in Indian factories. This includes requirements for proper ventilation, adequate lighting, safe drinking water, and periodic health check-ups for workers exposed to hazardous materials.
- **Example** – A chemical plant must follow the Act's regulations by installing ventilation systems to control airborne toxic substances. Regular maintenance of this system is required, ensuring worker health and adherence to legal standards.

3.3.4 Control of Substances Hazardous to Health (COSHH) – United Kingdom

- **Overview** – COSHH requires employers to manage risks associated with hazardous substances, including proper labeling, handling, storage, and disposal. Factories using dangerous chemicals must take extra precautions to protect workers and the environment.

- **Example** – In a paint manufacturing facility, the maintenance team must regularly inspect and maintain storage tanks to prevent chemical leaks or spills. All workers handling these chemicals need training, and protective equipment must be readily available.

3.3.5 Environmental Protection Act – Canada

- **Overview** – This Act requires Canadian factories to manage and minimize their environmental impact, focusing on waste reduction, emissions control, and pollution prevention.
- **Example** – A factory emitting industrial gases must maintain emissions filters and inspect them periodically to comply with environmental regulations. Non-compliance, such as exceeding permissible emission levels, can lead to fines and mandatory shutdowns.

3.3.6 Fire Services Act – Australia

- **Overview** – In Australia, the Fire Services Act requires factories to have fire safety plans, conduct fire drills, and maintain firefighting equipment.
- **Example** – In an automotive plant, fire extinguishers and sprinkler systems must be maintained regularly. The maintenance team is responsible for checking these systems to ensure they are operational and compliant with the Act.

3.3.7 Importance of Factory Legislation in Maintenance

- **Improves Workplace Safety** – Factory legislation reduces workplace accidents and injuries by enforcing safety standards, benefiting both workers and employers.
- **Ensures Equipment Reliability** – Maintenance requirements mandated by law prevent machinery from deteriorating, reducing downtime and production losses.
- **Reduces Environmental Impact** – Compliance with environmental laws ensures that factories manage waste and emissions responsibly, minimizing pollution and environmental harm.
- **Promotes Worker Welfare** – Legislation protects worker health, mandates breaks, and limits working hours, creating a balanced and supportive work environment.
- **Avoids Legal Consequences** – By adhering to factory legislation, companies avoid costly fines, litigation, and potential shutdowns due to non-compliance.
- **Improves Public Image** – Compliance with safety and environmental standards improves a company's reputation among employees, customers, and stakeholders.

3.3.8 Conclusion

Factory legislation is essential in maintaining safe, efficient, and ethical industrial operations. It guides maintenance practices, improves safety standards, and ensures legal compliance across various operational areas, including health, safety, equipment maintenance, and environmental protection. Examples from different countries show how these laws protect workers and the environment while enhancing factory productivity.

CHAPTER FOUR

4.0 ENSURING WORKPLACE SAFETY: THE PERMIT-TO-WORK SYSTEM AND HAZARD MANAGEMENT

4.1 PERMIT-TO-WORK SYSTEM: OVERVIEW AND EXAMPLES

A Permit-to-Work (PTW) system is a formal process that manages high-risk or potentially hazardous work activities in industrial environments. It involves an official document or "permit" authorizing certain maintenance or operational tasks only after identifying hazards and implementing necessary safety measures. This system ensures that everyone involved understands the risks, safety procedures, and scope of work, providing structured controls to protect workers, equipment, and facilities.

4.1.1 Key Elements of a Permit-to-Work System

1 **Hazard Identification** – Recognizing and analysing potential dangers associated with specific tasks, such as electrical hazards, chemical exposure, or confined spaces.

2 **Risk Assessment**: –Evaluating the risks associated with identified hazards and establishing measures to control or mitigate these risks.

3 **Approval Process** – Getting approvals from authorized personnel (such as supervisors or safety officers) who are accountable for verifying that all safety measures are in place.

4 **Documentation** – Recording the permit with essential information, including the type of work, date, time, location, personnel involved, safety measures, and duration.

5 **Monitoring and Compliance** – Ensuring that all workers comply with the permit's requirements and monitoring ongoing work to identify any emerging hazards or changes.

6 **Permit Closure** – Closing the permit only after confirming that the task is complete and the worksite has been restored to a safe condition.

4.1.2 Types of Work Requiring a Permit

1 **Hot Work** – Activities involving heat, flames, or sparks, such as welding or cutting, which carry a fire risk.

2 **Confined Space Entry** – Work in enclosed areas with limited access, such as tanks or silos, where there is a risk of asphyxiation or toxic gas exposure.

3 **Electrical Work** – Work on high-voltage equipment or circuits that poses an electrocution risk.

4 **Work at Height** – Tasks that involve working above ground, requiring fall prevention and protection measures.

5 **Handling Hazardous Substances** – Work with chemicals or other hazardous materials that requires specialized handling and protective equipment.

4.1.3 Examples of Permit-to-Work Systems

4.1.3.1 Oil Refinery - Hot Work Permit

Example – In an oil refinery, a maintenance team needs to repair a pipeline using welding. Since this is a high-risk activity that involves fire hazards, a hot work permit is required. The permit process includes shutting down nearby flammable processes, testing for the presence of explosive gases, and ensuring fire-fighting equipment is nearby. A supervisor signs off on the permit, allowing welding to proceed only once the work area is safe.

4.1.3.2 Confined Space Entry in a Wastewater Treatment Plant

Example – A wastewater treatment worker needs to enter a storage tank to inspect it for maintenance. The confined space entry permit includes testing for oxygen levels, assessing the presence of toxic gases, and ensuring a standby rescue team is ready. Only once these precautions are verified and approved by a safety officer is the worker allowed to enter the tank.

4.1.3.3 Electrical Work Permit in a Manufacturing Plant

Example – Electricians are tasked with replacing circuit breakers on a high-voltage system in a manufacturing plant. Before work begins, they obtain an electrical permit, ensuring power is isolated and locked out (lockout/tagout) to prevent accidental energizing. Only after a supervisor verifies these precautions does the team receive authorization to proceed.

4.1.3.4 Work at Height Permit for Wind Turbine Maintenance

Example – Technicians performing maintenance on wind turbines at high elevations require a working at height permit. The permit includes ensuring that harnesses, lanyards, and anchor points are inspected and set up correctly. It also requires a rescue plan in case of a fall. This permit is reviewed and signed off by the safety officer, allowing technicians to work at height safely.

4.1.4 Benefits of a Permit-to-Work System

1 **Enhanced Safety** – PTW systems protect workers from potential hazards, reducing the risk of accidents.

2 **Clear Communication** – With formal documentation and procedures, workers are fully aware of task-specific hazards, safety measures, and responsibilities.

3 **Regulatory Compliance** – Adhering to PTW systems helps organizations comply with industry standards and legal requirements for workplace safety.

4 **Accident Prevention** – By following strict controls, the PTW system reduces incidents resulting from hazardous activities, equipment malfunctions, or human error.

5 **Improved Accountability** – A formal permit system assigns clear responsibility to those approving and performing tasks, making it easier to track compliance and address safety concerns.

4.1.5 Challenges of Implementing a Permit-to-Work System

- **Administrative Burden** – The documentation process can be time-consuming and may slow down urgent tasks if not managed efficiently.
- **Lack of Compliance** – In cases where permits are not followed precisely, the system becomes less effective, potentially leading to unsafe work environments.
- ***Over-Reliance on Permits*** – Permits cannot replace practical safety measures. Workers should not view the permit as a complete substitute for physical precautions or professional judgement.

4.1.6 Conclusion

The Permit-to-Work system is essential for managing high-risk activities, ensuring that specific work environments remain safe for all involved. Examples from diverse industries, such as oil refineries, wastewater treatment, and manufacturing, demonstrate the system's versatility in safeguarding workers across various tasks. With its structured approach to hazard management, the PTW system helps reduce accidents and injuries while upholding industry safety standards and regulations.

4.2 SAFETY AND HAZARDS: OVERVIEW AND EXAMPLES

Safety and hazards in an industrial or workplace context encompass the identification, assessment, and management of risks that could potentially harm workers, equipment, or the environment. The goal of safety practices is to mitigate these risks through regulations, training, equipment, and processes that promote a safe working environment. Hazards are the potential sources of harm or adverse effects, while safety practices are the preventive measures taken to avoid accidents, injuries, and environmental damage.

4.2.1 Types of Hazards in the Workplace

- **Physical Hazards** – These include exposure to noise, heat, vibrations, and radiation that can physically harm workers. For instance, loud machinery can lead to hearing loss if proper hearing protection is not used.
- **Chemical Hazards** – These occur when workers handle or are exposed to hazardous substances, such as chemicals, gases, or toxic materials. Handling strong acids without appropriate personal protective equipment (PPE) can lead to skin burns and respiratory issues.
- **Biological Hazards** – Risks include exposure to bacteria, viruses, or mold, especially relevant in healthcare, laboratories, and agricultural settings. For example, handling contaminated medical waste without proper PPE could expose workers to infections.
- Ergonomic Hazards – These result from repetitive movements, poor posture, or inadequate workstation design, leading to musculoskeletal disorders. Prolonged, improper lifting techniques can cause back injuries in warehouse workers.
- **Electrical Hazards** – Risks associated with electrical currents, such as shocks, burns, or electrocution. Electricians working without de-energizing equipment are at risk of electric shocks.
- **Mechanical Hazards** – Result from machinery or moving parts that can cause injuries like cuts, crush injuries, or amputations. Unshielded gears or conveyor belts can trap clothing or body parts, leading to serious injuries.

4.2.2 Key Safety Practices to Mitigate Hazards

1 **Personal Protective Equipment (PPE)** – Using gloves, helmets, goggles, masks, and other gear to reduce exposure to hazards. For example, welders wear helmets and gloves to protect against sparks and heat.

2 **Safety Training and Education** – Ensuring employees are trained on hazard recognition, safe operating procedures, and emergency response actions. In a lab setting, workers are trained in proper chemical handling and spill procedures.

3 **Signage and Labels** – Placing visible signs and labels around the workplace to warn workers of hazardous areas and equipment. A "Caution: Wet Floor" sign helps prevent slips and falls.

4 **Emergency Response Plans** – Developing and practicing emergency procedures for situations like fires, spills, and medical emergencies. A factory with flammable materials would conduct fire drills and have extinguishers accessible.

5 **Engineering Controls** – Modifying equipment or processes to make them safer, such as using exhaust ventilation in areas where toxic fumes are present.

6 **Regular Maintenance and Inspections** – Scheduling inspections and maintenance for machinery and tools to prevent breakdowns and potential accidents.

4.2.3 Examples of Workplace Hazards and Safety Measures

4.2.3.1 Chemical Hazard in Manufacturing

- **Example** – In a chemical manufacturing plant, workers handle strong acids and bases that can cause skin burns or respiratory issues.
- **Safety Measure** – Workers are required to wear full-body protective suits, gloves, and face shields, and the plant has a specialized ventilation system to prevent fume inhalation. Training on handling spills and emergency wash stations are also provided.

4.2.3.2 Physical Hazard in Construction

- **Example** – Construction workers are exposed to loud machinery like jackhammers and drills that can lead to hearing damage.
- **Safety Measure** – Workers wear hearing protection, such as earplugs or earmuffs, and noise levels are regularly monitored. Additionally, signage is posted in high-noise areas to alert others to the noise hazard.

4.2.3.3 Biological Hazard in Healthcare

- **Example** – Healthcare professionals are at risk of contracting infectious diseases when handling patients or biological samples.
- **Safety Measure** – They are trained in infection control practices, including the use of gloves, gowns, and masks, and follow strict protocols for disposing of biohazardous waste to prevent contamination.

4.2.3.4 Electrical Hazard in Manufacturing

- **Example** – Technicians working on high-voltage electrical systems risk electrocution if they accidentally come into contact with live circuits.
- **Safety Measure** – Technicians follow lockout/tagout procedures to ensure equipment is de-energized before performing maintenance. Only authorized personnel are allowed to access electrical panels, and insulated tools are used for added protection.

4.2.3.5 Ergonomic Hazard in Office Setting

- **Example** – Office workers who sit for long periods or have poorly adjusted chairs and desks often experienced back pain or carpal tunnel syndrome.
- **Safety Measure** – Employers provide adjustable chairs, ergonomic keyboards, and encourage regular breaks to stretch or move. Workers are trained on proper posture and workstation arrangement to minimize strain.

4.2.3.6 Mechanical Hazard in Warehouse Operations

- **Example** – Workers in a warehouse environment may operate forklifts or work around moving conveyor belts, risking crush injuries.
- **Safety Measure** – Workers wear high-visibility vests, follow safety protocols when operating machinery, and work only in designated pathways. Emergency stop buttons are placed near conveyors to quickly stop operations if someone is in danger.

4.2.4 Importance of Addressing Safety and Hazards

1. **Prevention of Injuries and Fatalities** – Identifying hazards and implementing safety measures protect workers from accidents and severe injuries.
2. **Legal and Regulatory Compliance** – Ensuring compliance with occupational health and safety regulations reduces the risk of fines and legal action.
3. **Increased Productivity** – A safe workplace minimizes downtime and fosters a focused, efficient work environment.
4. **Employee Morale and Retention** – Showing a commitment to safety builds trust with employees, leading to improved morale and reduced turnover.

4.2.5 Conclusion

Managing safety and hazards is a vital aspect of workplace operations, requiring proactive measures and continuous monitoring to protect workers and maintain operational efficiency. By implementing thorough training, PPE usage, hazard identification, and emergency procedures, companies can significantly reduce the risks associated with workplace hazards. The examples demonstrate how specific industries manage diverse types of hazards, displaying the importance of tailored safety strategies across various work environments.

CHAPTER FIVE

5.0 WORKSHOP MANAGEMENT AND MAINTENANCE OPERATIONS

5.1 TYPES OF WORKSHOP AND REPAIR FACILITIES

Workshop and repair facilities play a vital role in maintaining and extending the life of equipment, machinery, and other assets in various industries. Each type of facility has a specialized setup to handle specific types of maintenance, repair, and troubleshooting work. These facilities are critical in sectors like manufacturing, transportation, automotive, construction, and electronics, where equipment and machinery need to be in optimal working order to ensure productivity, safety, and efficiency.

5.1.1 Mechanical Workshops

Mechanical workshops are designed for the maintenance and repair of mechanical equipment. These facilities have specialized tools and equipment to work on machinery parts, engines, and mechanical assemblies.

Example – In a manufacturing plant, a mechanical workshop is responsible for the upkeep and repair of assembly line machinery. It includes tools like lathes, drill presses, grinders, and hydraulic lifts for tasks like replacing worn-out machine parts, lubricating gears, or fixing motor issues.

5.1.2 Electrical Repair Workshops

Electrical repair workshops focus on maintaining and troubleshooting electrical equipment and systems. Technicians in these workshops work on wiring, circuits, motors, and electronic components.

Example – In a factory, an electrical repair workshop might service electric motors, generators, and control panels. It is equipped with multimeters, oscilloscopes, soldering stations, and circuit testers to diagnose and repair electrical faults or perform rewiring tasks.

5.1.3 Automotive Repair Facilities

Automotive repair facilities are specialized in servicing vehicles, including engine repair, transmission service, brake and suspension work, and body repair. They serve both light and heavy vehicles.

Example – A fleet maintenance workshop for a logistics company performs routine maintenance like oil changes, brake checks, and tire rotations. Heavy-duty equipment like hydraulic lifts and specialized diagnostic software helps technicians perform in-depth analysis and repairs on fleet trucks and vans.

5.1.4 Fabrication Workshops

Fabrication workshops are dedicated to creating, modifying, and repairing metal structures or parts. These facilities are equipped for welding, cutting, forming, and assembling metal components.

Example – In a construction company, a fabrication workshop produces custom metal fixtures, structural components, or repair parts for equipment. Equipment such as CNC machines, plasma cutters, and welding machines enables technicians to manufacture or repair complex metal parts.

5.1.5 Electronics Repair Workshops

Electronics repair workshops handle the maintenance and repair of electronic devices and systems, focusing on circuit boards, digital devices, and sensitive electronic components.

Example – An electronics repair workshop in a telecommunications company may fix damaged circuit boards, repair communication devices, and troubleshoot electronic control systems. Essential tools include circuit testers, anti-static workstations, and specialized software for diagnosing electronic faults.

5.1.6 Hydraulic and Pneumatic Repair Facilities

These facilities specialize in repairing hydraulic and pneumatic equipment, which are essential in various industries like manufacturing, construction, and transportation.

Example – In a mining operation, a hydraulic repair workshop may service hydraulic cylinders, pumps, and valves used in heavy mining equipment. The facility would have hydraulic test benches, fluid testing equipment, and high-pressure pumps to ensure that the hydraulic systems are functioning correctly.

5.1.7 Woodworking and Carpentry Workshops

Woodworking workshops handle the repair, fabrication, and maintenance of wooden structures and furniture. These workshops are often found in construction companies, furniture manufacturing, and maintenance departments of institutions like schools or hotels.

Example – A carpentry workshop in a hotel is responsible for repairing furniture, constructing new wooden fixtures, and maintaining the aesthetic quality of the premises. Tools include saws, sanders, lathes, and planers, allowing craftsmen to build or restore wooden furniture and fittings.

5.1.8 Welding and Metalworking Workshops

Welding and metalworking workshops perform welding, cutting, and metal joining operations. They are critical in repair and maintenance for industries where metal structures and equipment are used.

Example – In shipbuilding, a welding workshop repairs and maintains metal components of ships and marine equipment. It is equipped with various welding machines (MIG, TIG), gas cutters, and grinders to join or repair heavy metal sections, ensuring the structural integrity of vessels.

5.1.9 HVAC Repair Facilities

HVAC (Heating, Ventilation, and Air Conditioning) repair facilities specialize in maintaining and repairing climate control systems. These are essential in commercial buildings, manufacturing plants, and healthcare facilities.

Example: An HVAC repair facility in a large office complex maintains air conditioning units, heating systems, and ventilation ducts. Equipment includes vacuum pumps, refrigerant recycling machines, and leak detectors, which allow technicians to handle tasks such as replacing filters, sealing leaks, and calibrating thermostats.

5.1.10 Computer and IT Repair Facilities

Computer and IT repair facilities handle the repair, maintenance, and troubleshooting of computers, servers, networking equipment, and other digital systems.

Example – In a corporate IT department, an IT repair facility diagnoses and repairs issues related to computers, servers, and printers. Tools include anti-static equipment, data recovery software, and diagnostic kits, which enable technicians to troubleshoot hardware malfunctions, replace components, and perform software upgrades.

5.1.11 Laboratory and Precision Equipment Repair Facilities

These facilities service specialized laboratory equipment and precision instruments used in fields like healthcare, pharmaceuticals, and scientific research.

Example – A laboratory repair facility in a hospital is responsible for the calibration and maintenance of lab instruments like microscopes, centrifuges, and spectrometers. Technicians use precision calibration tools, cleaning solutions, and spare parts to ensure the accurate performance of medical instruments.

5.1.12 Facility Maintenance Workshops

Facility maintenance workshops are general-purpose workshops within an organization, responsible for the repair and upkeep of the building infrastructure and utilities.

Example – In a university, a facility maintenance workshop handles repairs in plumbing, lighting, HVAC, and general building maintenance. The workshop includes basic tools for plumbing, electrical work, and carpentry to handle day-to-day repairs and infrastructure maintenance.

5.1.13 Summary

Each workshop and repair facility is tailored to meet the specific needs of an industry or organization, ensuring that equipment, infrastructure, and assets are maintained efficiently. By employing specialized tools, techniques, and skilled technicians, these facilities are essential in reducing downtime, improving safety, and prolonging the lifespan of valuable assets.

5.2 MAINTENANCE POLICIES

Maintenance policies are formalized strategies that outline how maintenance activities are conducted within an organization. They serve as guidelines for maintaining equipment, infrastructure, and other assets, ensuring that assets remain in optimal working condition, productivity is maximized, and unplanned downtimes are minimized. Different industries implement specific maintenance policies based on the types of assets they operate, regulatory requirements, cost implications, and operational demands.

Here are some common types of maintenance policies, along with examples to illustrate their application:

5.2.1 Preventive Maintenance Policy

A preventive maintenance policy focuses on scheduled inspections, servicing, and routine repairs of equipment to prevent failures. This approach is based on regular intervals or usage, aimed at addressing wear and tear before it leads to significant breakdowns.

Example – In a manufacturing plant, preventive maintenance is scheduled for production machinery, where equipment is inspected, lubricated, and calibrated every three months. This ensures that the machines continue to operate efficiently, reduces the risk of breakdowns, and maintains the quality of production.

5.2.2 Predictive Maintenance Policy

A predictive maintenance policy uses real-time data to monitor the condition of equipment and predict when maintenance should be performed. This approach relies on technologies such as sensors, data analysis, and machine learning algorithms to provide accurate maintenance forecasts.

Example – In an oil and gas company, predictive maintenance sensors are attached to drill pumps to monitor their vibration levels, temperature, and pressure. If any of these indicators fall outside normal ranges, the system alerts the maintenance team to address potential issues before they result in a breakdown.

5.2.3 Reactive Maintenance Policy

Reactive maintenance, or "run-to-failure" policy, involves performing maintenance only after equipment has failed. This approach is generally used for non-critical equipment where the cost of downtime is low, or if the equipment's replacement is more economical than its regular maintenance.

Example – In a small food production company, a run-to-failure policy might apply to inexpensive equipment like conveyor belts or carts. If one of these items breaks, it is simply repaired or replaced rather than receiving ongoing scheduled maintenance.

5.2.4 Reliability-Centered Maintenance (RCM) Policy

Reliability-Centered Maintenance (RCM) is a policy that focuses on understanding and preserving the functions of critical systems within an organization. This policy determines the most cost-effective maintenance strategy for each asset based on its criticality and failure modes.

Example – In the aviation industry, RCM is used for critical components such as engines and landing gears. Through failure mode analysis and detailed risk assessments, the maintenance policy specifies high-frequency inspections for these parts to ensure passenger safety and compliance with aviation standards.

5.2.5 Condition-Based Maintenance (CBM) Policy

Condition-Based Maintenance involves conducting maintenance only when specific indicators show signs of decline. CBM relies on sensors, diagnostics, and visual inspections to determine when maintenance is required.

Example – In a data center, CBM is applied to cooling systems where temperature and humidity levels are constantly monitored. If the temperature within a server room begins to rise outside safe operating limits, maintenance technicians are alerted to check the HVAC system to prevent equipment overheating.

5.2.6 Deferred Maintenance Policy

A deferred maintenance policy is, when necessary, maintenance activities are delayed due to budget constraints, resource limitations, or strategic reasons. This policy is often used as a temporary measure to prioritize critical systems over fewer essential ones, though it can carry risks of higher costs and system failures over time.

Example – In a university setting, maintenance on non-critical buildings may be deferred to allocate more resources to main lecture halls and libraries. While deferred maintenance can save money short-term, it may lead to greater repair needs if left too long.

5.2.7 Shutdown Maintenance Policy

This policy involves performing maintenance only during planned shutdowns, usually for critical equipment that cannot be serviced while in operation. Shutdowns are scheduled when production is low, such as during holidays or weekends, to minimize disruption.

Example – In a steel manufacturing plant, shutdown maintenance is planned during holiday periods to conduct major overhauls on the blast furnace. Since the blast furnace cannot be safely repaired while operational, a complete shutdown is essential to address any necessary maintenance tasks and replace worn-out components.

5.2.8 Opportunity-Based Maintenance Policy

Opportunity-based maintenance takes advantage of production downtime to perform maintenance tasks. This policy is often used in industries where production schedules are flexible, allowing maintenance teams to address minor issues without a full shutdown.

Example – In a packaging plant, if one of the production lines stops for material replenishment, the maintenance team uses this downtime to check bearings, belts, and motors on the conveyor. This prevents potential issues from worsening and maximizes the use of idle time.

5.2.9 Corrective Maintenance Policy

Corrective maintenance is a policy focused on fixing defects or malfunctions after they have been identified. This policy is used for issues that do not necessarily halt operations immediately but require timely repairs to prevent escalation.

Example – In a hospital, if a wheelchair or an IV stand is found to be malfunctioning, corrective maintenance is conducted to repair or replace the equipment as soon as possible. Since patient care is a priority, such corrective actions ensure that essential equipment remains in working condition.

5.2.10 Proactive Maintenance Policy

Proactive maintenance is aimed at eliminating root causes of equipment failure to increase overall system reliability. By identifying and addressing the factors that lead to equipment wear or deterioration, this policy enhances the longevity and reliability of assets.

Example – In a mining operation, if repeated failures in conveyor belts are caused by dust accumulation, a proactive maintenance policy might involve installing better dust suppression systems and scheduling regular cleaning to address the root cause and prevent further belt failures.

5.2.11 Summary

Choosing the right maintenance policy is essential for balancing reliability, efficiency, and cost-effectiveness. Factors such as asset criticality, budget, risk, and operational demands guide the selection of appropriate maintenance policies. Implementing these policies helps organizations maximize asset life, reduce unexpected downtime, and maintain a safe and productive working environment.

5.3 ORGANIZATIONAL STRUCTURE AS A DECISION-MAKING CENTER

The organizational structure of a company greatly influences its decision-making process. The structure outlines how authority, responsibility, and communication flow within the organization, shaping how decisions are made, who makes them, and how quickly they can be implemented. Different types of organizational structures—ranging from hierarchical to flat, functional to matrix—offer varying advantages and are suitable for different types of decision-making needs, depending on the industry and operational objectives.

Below are common organizational structures and examples of how they function as decision-making centers:

5.3.1 Hierarchical Structure

A clear, top-down authority chain characterizes a hierarchical structure. Decisions in this structure flow from senior management at the top down to employees at lower levels. This structure is typically used in large organizations where control and accountability are crucial.

Example – In a manufacturing company, a hierarchical structure places the plant manager at the top, followed by department heads for production, quality control, and maintenance. When deciding on production quotas or resource allocations, decisions are made at the upper management level and communicated down to each department, ensuring standardized and cohesive implementation.

5.3.2 Functional Structure

In a functional structure, an organization is divided into specialized departments, each focusing on specific areas such as marketing, finance, or operations. This setup allows each department to make decisions based on their expertise, with the head of each function often contributing to the overall strategy.

Example – In a healthcare organization, departments like finance, human resources, and patient care operate independently with specialized functions. When a decision about budget reallocation for patient services needs to be made, the finance team assesses financial implications, and the patient care team evaluates service requirements. Decisions are often finalized by a committee or executive board based on departmental input, ensuring decisions are well-informed and functional needs are met.

5.3.3 Matrix Structure

A matrix structure blends elements of both functional and project-based structures, creating a dual reporting system. Employees report to both a functional manager and a project or product manager, encouraging collaboration across departments. This structure enhances decision-making flexibility, allowing project-based decisions while retaining departmental control.

Example – In an engineering firm, an employee may report to both the mechanical engineering department manager and the manager of a specific project they're assigned to. If the project team needs additional resources, the project manager can coordinate with the functional manager to make a quick decision, balancing project-specific needs with department resources. This approach enables faster and more flexible decision-making.

5.3.4 Flat Structure

In a flat structure, there are few or no levels of middle management between staff and executives. This structure is common in smaller companies and startups, where direct communication and quick decision-making are critical. Decisions are typically made collaboratively, involving team members directly.

Example – A tech startup with a flat structure may have teams that handle development, marketing, and customer support, all reporting directly to the CEO. If a decision is needed about launching a new product feature, each team quickly collaborates, evaluates customer needs, and gives direct input. This structure speeds up decision-making as there are minimal levels of approval required.

5.3.5 Divisional Structure

A divisional structure organizes a company based on products, projects, or geographic regions, with each division operating semi-independently. Each division has its own resources and decision-making authority, allowing them to respond quickly to specific market or project demands.

Example – In a multinational consumer goods company, separate divisions manage products like food, beauty, and health care. Each division has its own R&D, marketing, and finance teams and makes decisions tailored to its market. The food division might decide to introduce a new product based on consumer trends in one region, while the beauty division could make decisions on expanding its e-commerce platforms. This enables divisional autonomy, with faster and market-focused decisions.

5.3.6 Network Structure

In a network structure, an organization has a central hub that coordinates various independent entities or teams. This structure relies on partnerships and outsourcing and is often used by companies that collaborate with external suppliers, consultants, or remote teams.

Example – A fashion brand using a network structure outsources production to third-party manufacturers, logistics to distribution partners, and marketing to a separate agency. Decisions regarding inventory, logistics, or production schedules are made by coordinating with these external partners. This decentralized decision-making structure is beneficial for flexibility and scalability, though it requires strong coordination from the central hub.

5.3.7 The Role of Organizational Structure in Effective Decision-Making

Each structure has a distinct impact on how quickly and effectively decisions are made. For instance:

Hierarchical and Functional Structures tend to have slower decision-making processes due to the chain of command but benefit from standardized and controlled implementations.

Matrix and Network Structures offer flexibility, allowing for specialized decisions to be made by different areas or partners, though they may require strong communication to avoid conflicts.

Flat and Divisional Structures facilitate rapid decision-making and adaptability, which is especially beneficial in dynamic industries where fast response is critical.

5.3.8 Summary

In practice, companies often use a mix of these structures to leverage the advantages of each. An organization's choice of structure is influenced by its goals, industry, operational requirements, and the need for innovation versus control. Properly aligned organizational structures allow decision-making processes to operate efficiently, aligning with the company's strategic objectives.

5.4 STORE-KEEPING AND STOCK CONTROL SYSTEMS

Store-keeping and stock control are essential elements of inventory management, ensuring that materials, supplies, and products are stored and tracked efficiently to meet demand without overstocking. Effective store-keeping and stock control prevent production delays, reduce wastage, minimize costs, and support efficient operations. These systems typically involve record-keeping, stock monitoring, and inventory assessment and are essential across industries such as manufacturing, retail, and healthcare.

5.4.1 Key Elements of Store-Keeping

5.4.1.1 Storage and Organization

This involves arranging items in a systematic way within storage facilities, warehouses, or stockrooms, typically by category, frequency of use, or physical size.

Example – A car parts manufacturing company organizes its warehouse by categorizing items by type (e.g., brakes, electrical components) and size, with high-demand items placed at the front for easy access, minimizing time spent retrieving frequently used items.

5.4.1.2 Labeling and Identification

Items are labeled with unique codes or tags that make tracking and identification easy. Barcodes, RFID tags, or QR codes are commonly used to streamline the process.

Example – In a retail store, items have barcode labels. When scanned at checkout or for inventory updates, the system logs sales or stock adjustments, making it easy to identify what's in stock and what needs restocking.

5.4.1.3 Record Keeping

Records of items received, used, or disposed of are maintained to provide accurate stock data. This can be done manually, with spreadsheets, or via automated software systems.

Example – A hospital keeps track of medical supplies like syringes and bandages by logging each item received and used. Accurate records help avoid shortages of essential supplies and maintain sufficient stock levels.

5.4.2 Types of Stock Control Systems

5.4.2.1 Periodic Inventory System

In this system, physical counts of stock are done at regular intervals (weekly, monthly, quarterly). This approach can be manual or assisted by basic software.

Example – A small clothing store does a manual count at the end of each month to update its inventory records, tracking sales trends and identifying any missing or damaged items.

5.4.2.2 Perpetual Inventory System

A perpetual system continuously tracks inventory changes in real time through automated software linked to sales or usage data. Every transaction updates the stock count.

Example – In a supermarket, each time an item is scanned at checkout, the system immediately updates inventory levels. Managers can quickly see what items need replenishing without conducting physical counts.

5.4.2.3 Just-in-Time (JIT)

JIT aims to keep inventory levels low by only ordering stock as it's needed for production or sales. This reduces storage costs but requires accurate demand forecasting.

Example – A smartphone manufacturer orders specific components like processors and screens only when needed for production. JIT reduces warehouse storage requirements but demands precise coordination with suppliers.

5.4.2.4 Economic Order Quantity (EOQ)

EOQ is a stock control formula used to determine the optimal order quantity to minimize costs associated with ordering and holding stock. EOQ considers demand rates, ordering costs, and holding costs to optimize stock levels.

Example – A publishing company uses EOQ to decide how many copies of a popular book to order at a time, balancing the ordering and holding costs while ensuring there's enough stock to meet demand.

5.4.2.5 ABC Analysis

ABC Analysis categorizes inventory into three groups based on importance and value—"A" being high-value, "B" moderate-value, and "C" low-value items. This approach prioritizes resources for managing high-value items closely.

Example – In an electronics store, expensive items like laptops and smartphones are classified as "A" items, which require frequent tracking, while accessories like chargers or phone cases are "C" items that are tracked less rigorously.

5.4.3 Examples of Stock Control Software Systems

Modern stock control is often managed with inventory management software that provides real-time tracking, automatic updates, and data analytics. Examples include:

- ***SAP*** – Used by large enterprises, SAP provides an integrated inventory management solution with modules for tracking materials, controlling warehouses, and managing orders.
- ***QuickBooks*** – Small to medium-sized businesses use QuickBooks for tracking inventory and generating reports on stock usage, providing simple inventory tracking tools and sales tracking.
- ***Zoho Inventory*** – Zoho offers multi-channel inventory management, which is useful for businesses with online and in-store sales. It tracks stock in real time, handles orders, and integrates with other systems.

5.4.4 Benefits of Effective Store-Keeping and Stock Control

- ***Reduced Costs*** – Efficient stock control reduces holding costs by avoiding overstocking and storage of excess items.
- ***Minimized Waste*** – By ensuring items are used before they expire or become obsolete, waste is minimized, especially for perishable or time-sensitive stock.
- ***Improved Productivity*** – When items are organized and readily available, employees spend less time retrieving items, enhancing productivity and service delivery.
- ***Better Forecasting*** – Historical stock data provides insights into demand trends, helping companies forecast future needs and adjust order quantities accordingly.

5.4.5 Summary

In summary, well-managed store-keeping and stock control systems provide operational efficiency, reduce costs, and ensure availability, supporting the smooth functioning of various departments. They are crucial for businesses to keep inventory at optimal levels, prevent stockouts, and enhance overall productivity.

5.5 COSTING AND MAINTENANCE BUDGETS

Costing and budgeting for maintenance are critical in planning, tracking, and controlling maintenance expenses within an organization. Effective maintenance budgeting helps optimize asset performance, extend equipment lifespan, reduce downtime, and manage costs proactively. These budgets typically account for various types of maintenance activities—such as preventive, corrective, and predictive maintenance—and outline both fixed and variable expenses related to labor, materials, equipment, and services.

5.5.1 Components of Maintenance Costs

Labor Costs – This includes wages, overtime, benefits, and other related expenses for maintenance personnel and outsourced services.

Example – A manufacturing plant may employ full-time technicians to perform regular maintenance on machinery. The costs of their salaries, training, and any overtime required for unplanned maintenance are included in the labor budget.

Material and Spare Parts – This encompasses costs for parts, supplies, and consumables necessary for repairs, replacements, and general maintenance work.

Example – An airline sets aside funds to replace aircraft parts, such as landing gear and tires, which wear out over time. Keeping an inventory of these parts ensures they're readily available, reducing downtime and improving aircraft availability.

5.5.1.1 Equipment Costs

Costs associated with maintenance tools, vehicles, and specialized diagnostic equipment. It includes both purchase and rental expenses if equipment is rented for a specific maintenance task.

Example – A hospital's maintenance department may need specialized equipment to test and repair medical devices. This budget line item ensures that funds are available to maintain or rent equipment for essential tasks.

5.5.1.2 Contractor and Outsourced Services

Often, organizations outsource specialized maintenance tasks to external contractors, especially for tasks requiring technical expertise.

Example – A data center might hire external contractors to perform bi-annual HVAC system maintenance, essential to prevent overheating and equipment failure.

5.5.1.3 Utility and Facility Costs

Expenses related to energy, water, and other utilities are part of facility maintenance costs, especially for large organizations or manufacturing plants.

Example – In a food processing plant, the refrigeration system uses a substantial amount of electricity. A portion of the maintenance budget is allocated to keeping this system running efficiently.

5.5.1.4 Miscellaneous Costs

Includes costs like transportation, insurance, training for maintenance staff, and unforeseen expenses associated with emergency repairs or system upgrades.

5.5.2 Types of Maintenance Budgets

5.5.2.1 Zero-Based Budgeting

Every line item must be justified from zero, building the budget from scratch based on expected needs rather than historical spending.

Example – A government facility uses zero-based budgeting for its maintenance, requiring each department to justify why specific resources are necessary for the upcoming year.

5.5.2.2 Incremental Budgeting

In this approach, the previous year's budget is used as a baseline, with adjustments made for inflation, anticipated growth, or changes in maintenance demands.

Example – A university increases its maintenance budget annually by 5% to account for inflation and the aging of its buildings.

5.5.2.3 Flexible Budgeting

Adjusts maintenance budgets based on actual activity levels or operational changes. This approach is ideal for industries with fluctuating demands, such as manufacturing.

Example – A power plant's maintenance budget is flexible, allowing adjustments based on the season's energy demands, with additional funds allocated for higher-demand periods.

5.5.2.4 Fixed Budgeting

Sets a predetermined budget with limited room for adjustments, used for routine maintenance with minimal variability.

Example – An office building may allocate a fixed amount for quarterly HVAC servicing, as it anticipates stable costs for this service throughout the year.

5.5.3 Example of Maintenance Budget Allocation

Considering a pharmaceutical plant that operates under stringent regulatory requirements. The maintenance budget may be allocated as follows:

- ***Preventive Maintenance*** – 40% of the budget allocated to regular inspections, lubrication, and component replacements to ensure the plant's machines operate within safety standards.
- ***Corrective Maintenance*** – 25% reserved for unexpected breakdowns, repairs, and system troubleshooting, providing a buffer to address urgent repair needs.
- ***Predictive Maintenance*** – 20% of the budget for predictive maintenance, such as diagnostic software subscriptions and sensor replacements, to reduce the likelihood of unexpected failures.
- ***Training and Development*** – 5% allocated to train staff on new equipment and techniques, ensuring maintenance is carried out safely and effectively.
- ***Equipment and Spare Parts*** – 10% for essential spare parts and tools necessary to maintain the production line efficiently.

5.5.4 Benefits of Maintenance Budgeting

1 **Cost Control** – By planning ahead, organizations can manage and control maintenance costs, reducing overspending.

2 **Asset Reliability** – Regular budgeting for preventive and predictive maintenance ensures assets run efficiently and minimizes breakdowns.

3 **Long-Term Planning** – Well-defined maintenance budgets support long-term financial and asset management strategies.

4 **Enhanced Safety** – Budgeting for training and safety equipment reduces workplace hazards and supports a safe environment.

5.5.5 Importance of Tracking and Adjusting Maintenance Budgets

Once a budget is established, tracking and adjusting it based on actual expenses is essential. Comparing the budget to actual spending helps managers understand variances, make informed decisions, and optimize resources for future budgeting.

For example, if a retail store's maintenance budget consistently exceeds the allocated amount for corrective maintenance, it may indicate the need to allocate more funds toward preventive maintenance, thus reducing emergency repairs and lowering long-term costs.

5.5.6 Summary

In summary, effective costing and maintenance budgeting are crucial for optimal resource management, long-term asset reliability, and financial control across industries. By carefully analysing maintenance needs and adjusting budgets accordingly, organizations can achieve efficient, cost-effective maintenance programs.

CHAPTER SIX

6.0 INTEGRATED PLANT MAINTENANCE: SUPPORTING SERVICES, COST EFFICIENCY, PROACTIVE DESIGN, AND ESSENTIAL TOOLS FOR INDUSTRIAL OPERATIONS

6.1 USE OF MAINTENANCE AIDS: CHARTS AND GRAPHS

Maintenance aids, such as charts and graphs, are essential tools that support efficient maintenance management and decision-making. They simplify complex data, highlight trends, and assist in forecasting maintenance needs, allowing managers to make data-driven decisions that improve equipment reliability, reduce downtime, and optimize resource allocation.

6.1.1 Types of Maintenance Aids

6.1.1.1 Gantt Charts

Purpose – Gantt charts visually represent maintenance schedules over a specific timeline. They show the start and end dates of maintenance tasks and illustrate dependencies between tasks, helping managers track progress.

Example – A Gantt chart could be used to plan preventive maintenance for a production line over the course of a month, outlining tasks such as inspection, lubrication, and part replacements, along with the assigned personnel for each task.

Table 6.1: Sample of Preventive Maintenance Gantt Chart– E.g.; May 2025

Task	**Start Date**	**End Date**	**Assigned Personnel**
Inspection	May 1	May 2	Technician A
Lubrication	May 3	May 4	Technician B
Part Replacement	May 5	May 7	Technician C
Inspection	May 15	May 16	Technician A
Lubrication	May 17	May 18	Technician B
Part Replacement	May 20	May 22	Technician C

6.1.1.2 Pareto Charts

Purpose – Pareto charts help identify the most significant causes of maintenance issues, allowing teams to focus on the "vital few" that have the greatest impact on downtime or costs.

Example – In a manufacturing facility, a Pareto chart (Figure 6.1) might show that 80% of equipment failures are due to only 20% of the potential causes, such as specific components or environmental factors. Maintenance can then focus on addressing these high-impact causes.

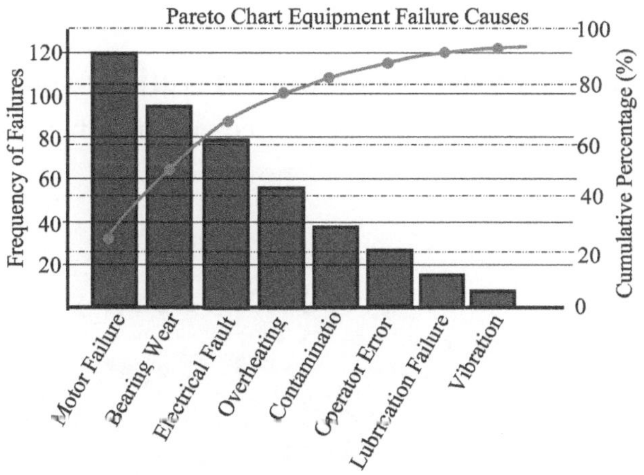

Figure 6.1: Sample of Pareto Chart

6.1.1.3 Bar Graphs and Histograms

Purpose – These graphs display the frequency of equipment breakdowns, parts usage, or other key metrics, offering insights into maintenance trends over time.

Example – A bar graph (Figure 6.2) could track the frequency of maintenance calls for different machines in a facility, helping identify which machines require more attention and resources.

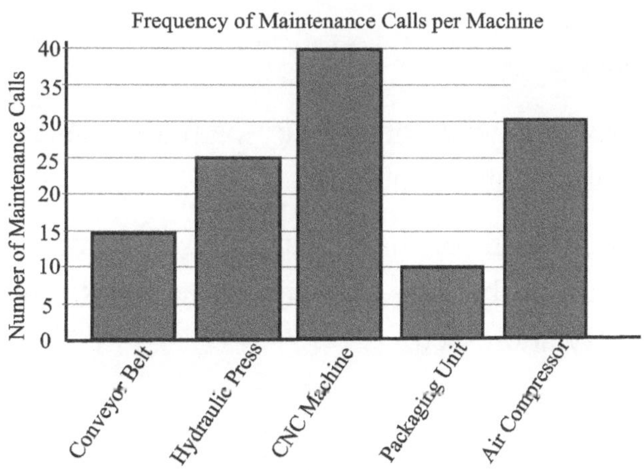

Figure 6.2: Sample of Bar Chart

6.1.1.4 Line Graphs

Purpose – Line graphs are useful for tracking trends over time, such as machine performance, repair costs, or downtime. They are particularly effective for showing gradual changes and fluctuations.

Example – A maintenance team might use a line graph to monitor monthly maintenance costs, highlighting any seasonal patterns or changes in expenses over time.

6.1.1.5 Pie Charts

Purpose – Pie charts represent the proportion of various maintenance activities, such as preventive, corrective, and predictive maintenance, within the overall maintenance budget or workload.

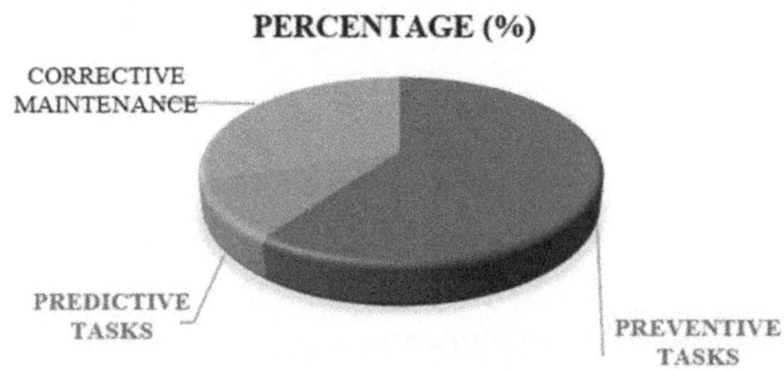

Figure 6.3: Sample of Pie Chart

Example – A pie chart (Figure 6.3) could display the allocation of maintenance hours, showing that 60% of time is spent on preventive tasks, 30% on corrective maintenance, and 10% on predictive tasks.

6.1.1.6 Control Charts

Purpose – Control charts track the consistency of maintenance processes and help identify variations that may signal potential problems.

Example – In a power plant, control charts may be used to monitor temperature and vibration levels of machinery, ensuring that any abnormal readings (indicating potential issues) as in Figure 6.4 are immediately addressed.

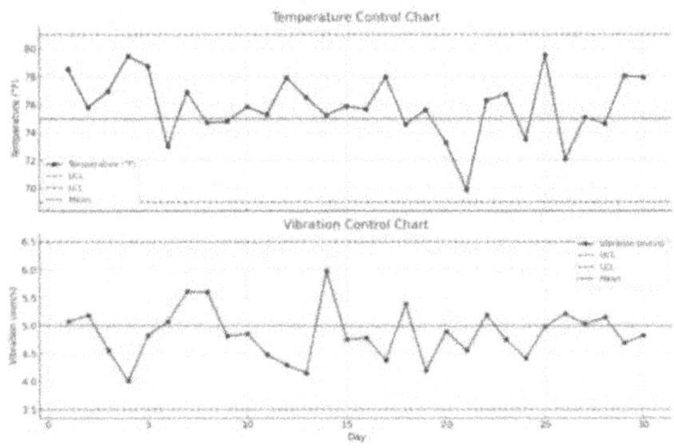

Figure 6.4: Example of Control Chart

6.1.1.7 Heat Maps

Purpose – Heat maps visually indicate areas of high activity or frequent breakdowns in a facility. They are helpful for identifying "hot spots" that need more attention.

Example – A heat map could show the parts of a facility where breakdowns occur most frequently, such as specific areas on a production line or sections of a warehouse as in Figure 6.5.

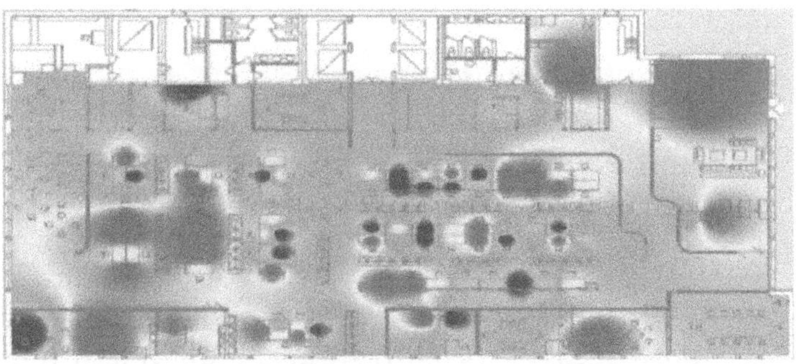

Figure 6.5: Sample of Heat Map

6.1.2 Benefits of Using Maintenance Aids

1 **Enhanced Decision-Making** – Maintenance charts and graphs make complex data easier to understand, allowing for faster and more accurate decision-making.

2 **Efficiency in Resource Allocation** – Visual aids highlight areas that require attention, helping allocate resources where they're most needed and reducing redundant work.

3 **Improved Predictive Maintenance** – By tracking historical data visually, trends that indicate potential future failures can be spotted and addressed proactively.

4 **Better Communication** – Charts and graphs make it easy to communicate maintenance performance and needs to stakeholders, ensuring everyone is informed and aligned.

6.1.3 Practical Examples of Maintenance Aids in Use

Example 1 – In a logistics company, a line graph showing average vehicle maintenance costs each quarter helps the maintenance manager anticipate high-cost periods and request additional budget allocations proactively.

Example 2 – In a food processing plant, a Pareto chart may reveal that frequent breakdowns of a mixer are due mainly to worn-out motor parts, suggesting the need for regular motor inspections and replacement schedules.

Example 3 – A pie chart used in a corporate building's maintenance department shows that 50% of their efforts are preventive, 30% corrective, and 20% predictive, enabling management to rebalance efforts if corrective tasks become too costly.

6.1.4 Summary

By providing insights and supporting analysis, maintenance aids such as charts and graphs enhance the overall maintenance strategy, making it easier to plan, track, and optimize activities across an organization. These tools are crucial for maintaining an efficient, reliable operation with reduced downtime and improved equipment life.

6.2 ECONOMIC MAINTENANCE

Economic maintenance focuses on balancing the cost of maintenance activities with the overall benefits they bring to an organization. The primary goal of economic maintenance is to minimize the total cost of equipment ownership, which includes initial investment, operating expenses, maintenance costs, and costs related to equipment downtime. By strategically managing these costs, organizations can maximize productivity, reduce downtime, and extend the life of their assets.

6.2.1 Key Concepts in Economic Maintenance

6.2.1.1 Cost-Benefit Analysis

Economic maintenance requires evaluating the costs and benefits of each maintenance action. The aim is to ensure that the benefits of maintaining an asset (such as reduced downtime and improved performance) outweigh the costs (such as labor, parts, and time spent).

Example – A factory may conduct a cost-benefit analysis on two options: performing preventive maintenance on a conveyor system every six months or waiting until a problem arises to make repairs. The analysis might show that regular preventive maintenance is more economical by preventing costly unplanned downtime.

6.2.1.2 Optimal Maintenance Frequency

Determining the optimal frequency of maintenance tasks is crucial in economic maintenance. Too frequent maintenance can result in excessive costs, while infrequent maintenance may lead to breakdowns.

Example – In a data center, regular cooling system maintenance is necessary to prevent overheating. By studying the failure history and cooling system wear patterns, the facility manager sets the maintenance frequency just right—often quarterly instead of monthly—thereby saving on labor and parts while keeping equipment reliable.

6.2.1.3 Asset Life Cycle Costing

Economic maintenance considers the entire life cycle of an asset, from acquisition to disposal. This approach ensures that maintenance strategies are aligned with the asset's expected lifespan, preventing overspending on assets nearing end-of-life.

Example – In a transportation company, if a truck is nearing the end of its life cycle, the company might choose to perform only minimal maintenance instead of costly overhauls. This allows the truck to operate safely until its scheduled replacement without unnecessary maintenance expenditures.

6.2.1.4 Maintenance Strategy Selection

Different types of maintenance strategies (e.g., preventive, predictive, or corrective) come with varying costs. Economic maintenance involves choosing the most cost-effective approach depending on the asset and operational environment.

Example – In a wind farm, predictive maintenance using condition monitoring (tracking parameters like vibration and temperature) is chosen over preventive maintenance. This method is more economical, as it predicts when parts may fail, allowing the replacement of only necessary parts rather than performing generalized maintenance on the whole system.

6.2.2 Examples of Economic Maintenance in Practice

6.2.2.1 Manufacturing Plant

In a manufacturing plant, machines are monitored for real-time performance indicators, such as vibration and energy consumption. Based on these parameters, maintenance actions are triggered only when needed, reducing both unnecessary downtime and labor costs.

6.2.2.2 Airline Industry

Airlines practice economic maintenance by planning major overhauls at set intervals (e.g., every 10,000 flight hours). Between these intervals, minor checks and predictive maintenance are performed to keep aircraft operating safely. This approach prevents costly unscheduled downtime while maintaining compliance with safety regulations.

6.2.2.3 Facilities Management

In a large commercial building, economic maintenance may involve investing in smart sensors for the HVAC system to optimize energy use. Maintenance is performed based on data from the sensors, which monitor air quality, system efficiency, and usage, rather than on a fixed schedule. This reduces the HVAC maintenance cost while ensuring consistent air quality and comfort for occupants.

6.2.2.4 Automotive Fleets

For a delivery company with a large vehicle fleet, economic maintenance entails scheduling vehicle maintenance at regular intervals that maximize the vehicles' reliability while minimizing downtime. By implementing predictive analytics on mileage and driving conditions, vehicles receive tailored maintenance based on usage patterns, optimizing cost and performance.

6.2.2 Benefits of Economic Maintenance

- **Cost Savings** – Reduces unnecessary maintenance expenses by aligning maintenance actions with actual equipment conditions.
- **Enhanced Asset Longevity** – Optimized maintenance schedules help extend the useful life of equipment, increasing return on investment.
- **Improved Productivity** – By reducing unexpected breakdowns, economic maintenance minimizes downtime and enhances operational efficiency.
- **Informed Decision-Making** – Economic maintenance strategies rely on data analysis, enabling more strategic and data-driven decisions.

6.2.3 Summary

In summary, economic maintenance prioritizes cost-effective maintenance practices that align with asset performance, longevity, and the operational needs of the business. By balancing maintenance costs with asset performance, organizations can achieve greater operational efficiency and a stronger bottom line.

6.4 DESIGNING OUT MAINTENANCE

"Designing out maintenance" (DOM) refers to the process of eliminating or minimizing the need for maintenance through careful design and engineering practices. This concept emphasizes creating products, systems, or infrastructure with durability and reliability in mind to reduce future repair, servicing, and upkeep. By minimizing maintenance needs at the design stage, organizations can reduce costs, improve efficiency, and increase asset lifespan, leading to significant operational savings.

6.4.1 Key Principles of Designing Out Maintenance

6.4.1.1 Use of Durable and Long-Lasting Materials

Selecting materials that are resistant to wear, corrosion, and fatigue can reduce maintenance needs over time. Materials that do not degrade easily in a particular environment minimize breakdowns and the need for frequent repairs.

Example: In the construction of offshore oil rigs, stainless steel or corrosion-resistant alloys are used for structural components that are exposed to saltwater to prevent rust, eliminating the need for frequent maintenance due to corrosion.

6.4.1.2 Incorporating Redundant or Backup Systems

Redundant systems are designed to take over automatically in the event of a primary system failure, reducing the need for immediate repairs.

Example: Data centers often include redundant cooling systems to prevent overheating. If one system fails, the backup kicks in, preventing equipment damage and reducing the urgency for immediate maintenance.

6.4.1.3 Streamlined Design and Simplification

Complex systems with many components are often more maintenance-intensive. Simplified designs with fewer moving parts, joints, or welds can reduce points of failure and make the system less susceptible to breakdowns.

Example: Electric vehicles (EVs) have fewer moving parts compared to traditional combustion engine vehicles. The simplified design reduces the need for frequent oil changes, spark plug replacements, and other routine maintenance associated with conventional engines.

6.4.1.4 Self-Cleaning or Self-Lubricating Features

Incorporating self-cleaning or self-lubricating mechanisms can significantly reduce maintenance requirements by reducing the need for manual interventions.

Example: Some conveyor systems are designed with self-lubricating belts that reduce the need for frequent lubrication, minimizing maintenance time and the risk of system wear.

6.4.1.5 Accessible and Modular Design

Designing components in a modular, easily accessible way allows for quick repairs and easy replacement. While it doesn't eliminate maintenance, it can reduce the cost and time required for maintenance tasks.

Example: In commercial aircraft design, engineers ensure critical parts like engines and landing gear are easily accessible, reducing the time and complexity involved in servicing.

6.4.2 Examples of Designing Out Maintenance

6.4.2.1 Wind Turbines

Wind turbines are often located in remote, hard-to-reach areas, making maintenance challenging and costly. To reduce the need for regular maintenance, manufacturers incorporate self-lubricating bearings, durable materials, and remote monitoring systems that detect issues early. The result is a wind turbine with fewer maintenance needs and reduced operational costs.

6.4.2.2 Self-Cleaning Windows in Skyscrapers

Many modern skyscrapers use self-cleaning glass with a special coating that breaks down dirt when exposed to sunlight and allows rain to wash the surface clean. This technology minimizes the need for frequent manual window washing, especially on higher floors, reducing maintenance costs and enhancing safety.

6.4.2.3 Anti-Fouling Paint on Marine Vessels

Marine vessels, such as cargo ships and oil tankers, often use special anti-fouling paint on their hulls to prevent algae and barnacle buildup. This reduces the need for routine hull cleaning and minimizes drag, leading to better fuel efficiency and lower maintenance costs.

6.4.2.4 Maintenance-Free Batteries

Lead-acid batteries, commonly used in vehicles and other equipment, often require regular water refills. In contrast, maintenance-free batteries use a sealed design that eliminates the need for regular topping up, significantly reducing maintenance for users.

6.4.2.5 Railway Tracks with High-Durability Alloy

In high-traffic railway lines, tracks can be subjected to wear from continuous use. By using high-durability alloys or composites, engineers can extend the life of these tracks, reduce the frequency of maintenance and minimize downtime for track replacement.

6.4.3 Benefits of Designing Out Maintenance

1 ***Cost Savings*** – Reducing maintenance needs through design decreases labor and part costs over the asset's life cycle.

2 ***Increased Operational Efficiency*** – Fewer maintenance interruptions allow for more continuous operation and productivity.

3 ***Enhanced Safety*** – Reducing the frequency and complexity of maintenance activities reduces risks associated with equipment failure or servicing in hazardous environments.

4 ***Sustainability*** – By reducing the resources needed for maintenance (such as energy, lubricants, and spare parts), DOM can contribute to more sustainable operations.

6.4.4 Challenges in Designing Out Maintenance

Initial Investment – Higher upfront costs are often associated with selecting durable materials and integrating advanced technologies to reduce maintenance needs.

Design Limitations – Certain equipment or environments may impose constraints on design, limiting the feasibility of fully eliminating maintenance needs.

Evolving Technologies – Rapid advancements may render current "maintenance-free" designs obsolete, necessitating updates or redesigns.

6.4.5 Summary

In summary, designing out maintenance is a proactive approach that addresses maintenance challenges at the design stage, focusing on durability, simplicity, and self-sufficiency. By investing in maintenance-free or reduced-maintenance designs, organizations can achieve long-term cost savings, enhance operational efficiency, and foster safer work environments. Examples like self-lubricating machinery, anti-corrosion materials, and modular designs demonstrate how effective planning can minimize the need for repairs, benefiting industries ranging from manufacturing to transportation.

6.5 INDUSTRIAL PLANT: DEFINITION, FUNCTIONS, AND EXAMPLES

An industrial plant refers to a large facility or building where goods are produced or processed on a large-scale using machinery, equipment, and human labor. These plants are crucial to various industries, from manufacturing and energy production to chemical processing and pharmaceuticals. Industrial plants typically involve complex operations and require regular maintenance, skilled personnel, and advanced technology to ensure smooth functioning.

6.5.1 Key Features of an Industrial Plant

6.5.1.1 Production Process

The primary purpose of an industrial plant is to convert raw materials into finished goods. The production process is often automated and involves various stages, such as assembling, machining, packaging, or chemical processing.

6.5.1.2 Machinery and Equipment

Industrial plants are equipped with heavy machinery that performs specific tasks essential to production. Examples include conveyor systems, furnaces, cutting machines, and robotics.

6.5.1.3 Labor Force

Skilled and unskilled workers operate the machinery, monitor processes, and ensure the plant operates efficiently. Engineers, technicians, and machine operators form the backbone of the workforce.

6.5.1.4 Maintenance and Safety

Industrial plants require regular maintenance to ensure all equipment is functioning efficiently and to prevent breakdowns that can halt production. Safety is also paramount due to the risks associated with heavy machinery, chemicals, or electrical systems.

6.5.2 Types of Industrial Plants with Examples

6.5.2.1 *Manufacturing Plant*

A manufacturing plant produces tangible products, often involving a sequence of complex processes to turn raw materials into finished goods. These plants vary greatly depending on the products they produce.

Example – An automobile manufacturing plant assembles various components like engines, body parts, and electronics into finished vehicles. Companies like Toyota or General Motors operate large plants where robotics, conveyor belts, and skilled labor work together to produce thousands of cars daily.

Example – A textile manufacturing plant produces fabrics and clothing from raw materials like cotton, wool, or synthetic fibers. The production involves weaving, dyeing, and stitching machinery to create different types of garments.

6.5.2.2 *Power Generation Plant*

Power generation plants convert energy from different sources into electricity, which is then distributed to homes, businesses, and industries.

Example – A coal-fired power plant burns coal to generate steam, which drives turbines connected to electricity generators. These plants are often large and equipped with systems to control emissions and reduce environmental impact.

Example – A nuclear power plant uses nuclear fission to generate heat, which is then used to produce steam that drives turbines. These plants require rigorous safety standards due to the potential risks of radiation.

6.5.2.3 *Chemical Processing Plant*

Chemical processing plants produce chemicals for various industries, including pharmaceuticals, agriculture, and food production.

Example – A petrochemical plant processes raw petroleum into products such as gasoline, plastics, and chemicals. Refineries are large, complex plants that use distillation and other methods to extract useful chemicals from crude oil.

Example – A pharmaceutical plant produces medicines, including pills, vaccines, and syrups. These plants are equipped with clean rooms, mixing equipment, and packaging machines to ensure that products meet strict safety and quality standards.

6.5.2.4 *Food Processing Plant*

Food processing plants transform raw ingredients into packaged food products. These plants are critical to ensuring food safety, quality, and shelf-life.

Example – A dairy processing plant turns raw milk into products like cheese, butter, and yogurt. The plant uses equipment such as pasteurizers, homogenizers, and packaging machines to produce dairy products on a large scale.

Example – A meat processing plant processes livestock into meat products like sausages, steaks, and ground meat. These plants involve slaughtering, cleaning, and packaging, with strict hygiene protocols to ensure food safety.

6.5.2.5 Mining Plant

Mining plants extract and process minerals and ores from the earth, which are then used in industries such as construction, electronics, and energy.

Example – A gold mining plant extracts gold ore from mines and processes it into pure gold. The plant uses techniques like crushing, grinding, and chemical leaching to separate gold from the ore.

Example – A coal mining plant processes extracted coal for use in power generation and industrial applications. The plant typically includes equipment for sorting, washing, and sizing the coal before it is sent to customers.

6.5.3 Importance of Maintenance in Industrial Plants

Industrial plants require regular maintenance to ensure the smooth operation of machinery and to prevent breakdowns that can halt production. Preventive maintenance schedules help in detecting issues early, reducing the risk of costly downtime. Additionally, emergency maintenance teams are often on-site to fix unexpected breakdowns.

Example – In an automobile plant, regular maintenance of the assembly line robots is crucial. Any malfunction in the robotic arms can lead to production delays, affecting the delivery schedule and increasing costs.

6.5.4 Safety and Environmental Concerns in Industrial Plants

Industrial plants often deal with hazardous materials or conditions, such as high temperatures, toxic chemicals, or heavy machinery. Safety protocols are strictly enforced to protect workers, and environmental regulations are followed to minimize pollution.

Example – In a chemical processing plant, employees wear protective gear, such as gloves, masks, and safety glasses, to avoid exposure to toxic chemicals. The plant may also have systems in place to control emissions and reduce environmental impact.

6.5.5 Summary

In summary, industrial plants are the backbone of various industries, converting raw materials into finished goods, energy, or chemicals that drive the global economy. The complexity of operations in these plants requires skilled labor, advanced machinery, and strict maintenance and safety protocols. Whether it's a manufacturing, chemical, or power generation plant, the role of industrial plants in modern society is indispensable.

6.6 IDENTIFYING BASIC TOOLS – TYPES AND EXAMPLES

In various industries, the use of basic tools is essential for performing tasks such as repairs, maintenance, construction, and production. Basic tools can be divided into several categories, depending on the nature of work they are used for. These tools are vital to efficiently perform tasks, and knowing their proper use is important for safety, productivity, and quality. Here's an overview of different categories of basic tools along with examples:

Hand tools are operated manually and are the most fundamental tools used in nearly all types of work, from construction to repair and maintenance.

6.6.1 Hand Tools

- **Hammer** – Used to drive nails into wood, break objects, or shape metal.

  Example – A claw hammer is often used in carpentry to drive and remove nails.

- **Screwdrivers** – These are used to drive or remove screws.

  Example – A flat-head screwdriver can be used for slot-type screws, while a Phillips screwdriver is used for cross-type screws.

- **Wrenches** – Used to tighten or loosen nuts and bolts.

  Example – A combination wrench has one open-end and one box-end, providing flexibility in tightening bolts.

- **Pliers** – Multi-functional tools used for gripping, twisting, or cutting wires.

  Example – Needle-nose pliers are often used in electronics to handle small components.

- **Measuring Tape** – Used to measure distances and sizes accurately.

  Example – A 25-foot retractable measuring tape is commonly used in construction projects to measure materials.

6.6.2 Cutting Tools

Cutting tools are designed to cut materials like wood, metal, plastic, and wires.

- ***Handsaw*** – A tool used to cut wood or other materials manually.

  Example – A crosscut saw is used to cut across the grain of wood.

- ***Utility Knife*** A tool with a retractable blade used for cutting thin materials.

  Example – Utility knives are often used in warehouses for cutting packaging materials.

- ***Wire Cutter*** – Used to cut electrical wires or cables.

Example – Diagonal cutters can easily cut through wire during electrical installation.

6.6.3 Power Tools

Power tools are electrically powered and used for tasks that require more force than hand tools can provide.

- ***Drill*** – Used to make holes in materials like wood, metal, or concrete.
- ***Example*** – A cordless electric drill is versatile for home improvement projects.
- ***Circular Saw*** – A power saw used for cutting wood and other materials in straight lines.

  Example – A 7-inch circular saw is often used in construction to cut plywood and framing materials.
- ***Angle Grinder*** – A tool used for cutting, grinding, or polishing materials.

  Example – In metalworking, an angle grinder is used to grind down sharp edges or cut metal bars.

6.6.4 Precision Tools

Precision tools are used to perform detailed and delicate tasks, often in electronics, watchmaking, and small assembly processes.

- ***Caliper*** – Used for measuring the distance between two opposite sides of an object.

  Example – A vernier caliper is commonly used in machining to measure the thickness of metal parts accurately.
- ***Micrometer*** – Provides precise measurements of small objects such as wire or thin metal sheets.

  Example – A micrometer can measure thicknesses with an accuracy of up to 0.001 millimeters.
- ***Torque Wrench*** – Used to apply a specific torque to nuts and bolts, ensuring they are tightened to the correct tension.

  Example – Used in the automotive industry for tightening car engine bolts to precise specifications.

6.6.5 Marking and Measuring Tools

Marking and measuring tools help in planning and ensuring precision before cutting or assembling materials.

- ***Chalk Line*** – Used for marking long, straight lines on surfaces.

  Example – Chalk lines are used in construction to mark cutting lines for large surfaces such as floors or walls.

- ***Spirit Level*** – A tool used to check if surfaces are perfectly horizontal (level) or vertical (plumb).

  Example – A carpenter uses a spirit level to ensure that shelves are level when mounted on a wall.

6.6.6 Maintenance Tools

Maintenance tasks, especially in mechanical and electrical work, require specific tools for routine checks and repairs.

- ***Multimeter*** – Used to measure electrical properties like voltage, current, and resistance.

  Example – An electrician uses a multimeter to check the voltage level in electrical circuits.

- ***Pipe Wrench*** – A wrench used to tighten or loosen pipes and other cylindrical objects.

  Example – Plumbers use pipe wrenches to install or repair plumbing pipes.

- ***Grease Gun*** – A tool used to apply lubricant (grease) to moving mechanical parts.

  Example – In automotive maintenance, a grease gun is used to lubricate the joints in the car's suspension system.

6.6.7 Safety Tools

Safety tools are crucial in environments where workers are exposed to risks like electrical shock, falling objects, or chemical hazards.

- **Safety Glasses** – Protects eyes from debris, chemical splashes, and other hazards.

  Example – Workers in manufacturing plants wear safety glasses while operating cutting machines.

- **Hard Hat** – Protects the head from impact and falling objects.

  Example – Construction workers wear hard hats on-site to protect themselves from debris.

- **Ear Protection (Earplugs or Earmuffs)** – Reduces noise exposure in loud environments.

- ***Example*** – Factory workers in loud machinery environments wear earmuffs to prevent hearing damage.

6.6.8 Conclusion

Identifying and knowing how to use basic tools is fundamental in any industry that involves mechanical, construction, electrical, or manufacturing tasks. Whether it's a simple hand tool like a screwdriver or an advanced power tool like an electric drill, the proper use of tools increases efficiency and safety.

CHAPTER SEVEN

7.0 COMPREHENSIVE PLANT INFRASTRUCTURE: INSTALLATION, STRUCTURAL DESIGN, BUILDING TYPES, ALIGNMENT, AND INSTRUMENTATION IN MAINTENANCE SERVICES

7.1 PLANT INSTALLATION

7.1.1 Installation Procedure: Definition and Importance

The installation procedure refers to the step-by-step process of setting up equipment, systems, or machinery in a specific location to ensure they are operational and ready for use. Proper installation is crucial for ensuring the efficient functioning of equipment, minimizing risks, and maximizing safety and performance. Whether it's the installation of a simple appliance or a complex industrial machine, following a structured procedure is essential to avoid malfunction, damage, or unsafe conditions.

7.1.2 Key Steps in Installation Procedure

7.1.2.1 Preparation and Planning

The first step in any installation is thorough preparation. This involves understanding the equipment's specifications, preparing the installation site, and ensuring that all required tools, materials, and personnel are available.

Example – In the installation of a commercial HVAC system, preparation includes verifying the system's capacity, checking the building's layout for ductwork, and ensuring that power sources are available.

7.1.2.2 Site Inspection

Before the actual installation begins, a site inspection is conducted to ensure that the area meets the equipment's requirements, such as space, ventilation, and structural integrity.

Example – During the installation of a large industrial machine, the installation team inspects the floor to ensure it can support the weight of the machine and checks for any obstructions like pipes or cables.

7.1.2.3 Unpacking and Inspection of Equipment

Equipment should be unpacked and inspected for any damage that may have occurred during shipping or storage. Damaged components should be reported and replaced before proceeding with the installation.

Example – When installing a new server in a data center, the IT team will unpack the server, inspect it for any signs of damage, and ensure that all components (e.g., hard drives, power supplies) are included and functional.

7.1.2.4 Positioning and Mounting

Once the equipment has been inspected, it is positioned in the designated location. This may involve using mounting brackets, anchors, or leveling tools to ensure the equipment is stable and secure.

Example – In the installation of solar panels on a roof, the panels must be carefully positioned to maximize sun exposure, and then securely mounted to the roof using brackets and fasteners.

7.1.2.5 Connection to Power Supply

After the equipment is positioned, it needs to be connected to the appropriate power source, ensuring that the voltage, amperage, and phase match the equipment's requirements.

Example – For the installation of a heavy-duty industrial compressor, the installation team must ensure that it is connected to the correct voltage and that the electrical circuits can handle the power load.

7.1.2.6 Connection of Auxiliary Systems

In many installations, the equipment must be connected to additional systems such as water supply lines, gas, or data networks. Ensuring proper connections is critical for the system's functionality.

Example – In the installation of a CNC machine in a factory, the machine must be connected not only to power but also to air compressors (for pneumatic tools), water cooling systems, and a network for data transfer.

7.1.2.7 Calibration and Testing

Once installed, the equipment needs to be calibrated to ensure it functions correctly according to the manufacturer's specifications. Testing is also conducted to verify that all systems are working properly.

Example – After installing a medical imaging machine in a hospital, technicians calibrate the device to ensure accurate imaging, and conduct test scans to verify that it operates within the required parameters.

7.1.2.8 Final Inspection and Handover

After installation and testing, a final inspection is conducted to ensure everything is working correctly, and the installation site is cleaned up. The equipment is then handed over to the user, along with any necessary documentation such as manuals and warranties.

Example – When installing a commercial-grade kitchen appliance, the technician performs a final check of all connections and components, cleans the area, and hands over the instruction manual to the kitchen staff.

7.1.3 Examples of Installation Procedures in Various Industries

7.1.3.1 Manufacturing Industry

In manufacturing, installation often involves complex machinery that must be integrated into existing production lines. This requires precision and careful attention to how the new equipment will interact with the systems already in place.

Example – Installing a robotic arm in an automobile manufacturing plant involves precise calibration of the arm's movements to ensure it fits seamlessly into the assembly process, minimizing errors and downtime.

7.1.3.2 IT Industry

In the IT industry, installation procedures typically involve setting up hardware such as servers, routers, or workstations, along with configuring software to meet the organization's needs.

Example – Installing a new server in a data center involves configuring the server with the appropriate operating system, connecting it to the network, and testing its performance to ensure it can handle the expected workload.

7.1.3.3 Construction Industry

The construction industry frequently involves the installation of large systems such as plumbing, HVAC, and electrical systems in buildings. Each system has its own specific installation requirements that must be followed to ensure safety and compliance with building codes.

Example – Installing a central air conditioning system in a commercial building involves connecting the unit to the electrical system, ductwork, and thermostat, and testing it for cooling efficiency and air distribution.

7.1.3.4 Telecommunications Industry

In telecommunications, installation often involves setting up towers, antennas, or data lines to enable communication services. Proper installation is vital for signal strength and coverage.

Example – Installing a telecommunications tower involves erecting the structure, mounting antennas, and connecting the tower to the network. Signal tests are conducted afterward to ensure coverage in the area.

7.1.4 Importance of Following the Installation Procedure

- ***Safety*** – Following a structured installation procedure ensures that equipment is installed safely, minimizing risks of injury or accidents.
- ***Efficiency*** – A proper installation ensures the equipment functions at peak efficiency, reducing the likelihood of future malfunctions or downtime.
- ***Compliance*** – Many industries are governed by regulations that require installations to meet specific standards. Failing to follow the correct procedure can result in non-compliance and legal issues.
- ***Longevity*** – Proper installation can extend the lifespan of the equipment by preventing unnecessary wear and tear caused by incorrect setup.

7.1.5 Summary

In summary, the installation procedure is a vital process across industries, ensuring that equipment and systems are set up correctly and safely. Following a clear, step-by-step process—from preparation and site inspection to final testing—ensures that the equipment functions efficiently and minimizes future maintenance issues. Whether it's in manufacturing, telecommunications, or construction, the importance of a proper installation cannot be overstated.

7.2 DESIGN CONSIDERATIONS FOR CONCRETE FLOORS

When designing concrete floors, various factors need to be considered to ensure durability, safety, aesthetics, and suitability for the intended purpose. Concrete floors are used in diverse applications, ranging from industrial and commercial buildings to residential homes. The design must address the load-bearing capacity, type of use, environmental conditions, and future maintenance needs.

7.2.1 Load-Bearing Capacity

Concrete floors must be designed to withstand the loads they will bear. The type and intensity of load depend on the specific application—whether it's a warehouse, parking garage, industrial facility, or residential building. The designer must consider both the static (permanent) loads, such as machinery or furniture, and dynamic (moving) loads, such as vehicles or foot traffic.

Example – In a warehouse where forklifts and heavy goods are constantly moving, the concrete floor needs to be reinforced and thick enough to withstand significant load-bearing without cracking or deteriorating.

7.2.2 Durability and Strength

Concrete floors should be durable enough to last for years without requiring major repairs. This involves selecting the right concrete mix with appropriate strength, and in some cases, incorporating reinforcement like steel bars or mesh. The strength of the concrete mix is measured in terms of its compressive strength (measured in MPa or PSI). The higher the strength, the better the floor will perform under heavy loads.

Example – In an industrial facility where machinery exerts heavy, concentrated loads on the floor, high-strength concrete with reinforced steel bars may be necessary to prevent wear and cracking.

7.2.3 Type of Reinforcement

Reinforcement strengthens the concrete floor, helping to prevent cracking and improve its load-bearing capacity. Common reinforcement materials include steel bars (rebar), welded wire mesh, and fiber reinforcement. The type and amount of reinforcement depend on the expected load and stress on the floor.

Example – A parking garage floor might require steel reinforcement to support the weight of parked cars and the movement of vehicles, ensuring the floor doesn't crack under the constant load.

7.2.4 Control of Shrinkage and Cracking

Concrete tends to shrink as it dries, which can cause cracks to develop. Proper design must include strategies to minimize shrinkage and cracking. This involves placing control joints at regular intervals to control where cracks may occur and allowing for movement as the concrete expands and contracts due to temperature changes.

Example – In a large commercial building, control joints are cut into the concrete slab at specified distances to help guide the natural cracking process, preventing unsightly or problematic cracks from forming unpredictably.

7.2.5 Surface Finish

The surface finish of a concrete floor depends on its intended use. Some floors require a smooth, polished finish for aesthetic or functional reasons, while others may need a rough texture for slip resistance. For decorative purposes, concrete can be stained, polished, or patterned to enhance its appearance.

Example – In showrooms or offices, polished concrete floors with a high-gloss finish are often chosen for their aesthetic appeal, ease of cleaning, and durability, whereas in outdoor walkways, a textured finish is used to provide slip resistance.

7.2.6 Moisture Control and Waterproofing

Concrete floors, especially those in basements or ground-level applications, need proper moisture control to prevent water damage, mold growth, and degradation over time. This can involve installing vapor barriers, applying waterproof coatings, or using special moisture-resistant concrete mixes.

Example – In a basement installation, a vapor barrier beneath the concrete slab helps prevent moisture from seeping through the floor, which could cause damage to the structure or flooring materials placed on top.

7.2.7 Thermal Performance

Concrete has excellent thermal mass, meaning it can absorb, store, and release heat slowly, which can be an advantage in controlling indoor temperatures. Designers may consider using the floor's thermal properties to improve energy efficiency, especially in climates where temperature control is crucial.

Example – In sustainable building designs, concrete floors can be used as part of a passive solar heating strategy. During the day, the concrete absorbs heat from the sun, and at night, it releases the heat to keep the indoor environment warm.

7.2.8 Impact Resistance

Concrete floors in certain settings must be designed to resist impact damage from heavy objects or equipment being dropped on them. This is particularly important in industrial facilities where materials may be moved or stored.

Example – In a manufacturing facility, the floor should be designed to handle the impact of heavy tools or parts being dropped without suffering major cracks or dents.

7.2.9 Chemical Resistance

In environments where the floor will be exposed to chemicals, such as industrial facilities or laboratories, it is essential to use concrete that has been treated or coated with a chemical-resistant sealant. This helps prevent the concrete from corroding or deteriorating due to chemical spills.

Example – In a chemical processing plant, the concrete floor might be coated with a protective epoxy layer to resist corrosive chemicals and ensure the floor maintains its integrity over time.

7.2.10 Environmental Conditions

Environmental factors such as freeze-thaw cycles, humidity, and exposure to corrosive elements must be considered during the design process. In areas prone to freezing, concrete floors must be designed to resist cracking due to expansion and contraction.

Example – In regions with extreme temperature fluctuations, the concrete mix may include air-entraining agents to create tiny air pockets within the concrete, allowing it to expand and contract without cracking during freeze-thaw cycles.

7.2.11 Maintenance Requirements

Designers should consider the long-term maintenance needs of the concrete floor. Floors in high-traffic areas or heavy-duty applications may require more frequent cleaning, sealing, or resurfacing to maintain their appearance and performance.

Example: In a shopping mall, regular cleaning and resealing may be necessary to keep polished concrete floors looking their best while also protecting them from wear and tear caused by constant foot traffic.

7.2.12 Summary

In summary, designing concrete floors requires careful consideration of factors like load-bearing capacity, reinforcement, surface finish, moisture control, and environmental conditions. Each application demands specific attention to these design aspects to ensure the floor's functionality, durability, and safety. Examples from industrial, commercial, and residential settings illustrate how different design considerations play a role in creating effective and long-lasting concrete floors.

7.3 FOUNDATION AND FOOTING

Foundations and footings are essential components of any structure, as they provide the necessary support to transfer the building's load to the ground. Properly designed and constructed foundations ensure the stability and safety of a building, preventing settlement, cracks, and structural failure. Both foundation and footing must account for factors such as soil conditions, building load, and environmental conditions to achieve long-term durability.

7.3.1 Definition of Foundation and Footing

7.3.1.1 Foundation

The foundation is the lower part of a building structure that transfers its load to the earth and supports it. The foundation ensures that the building remains stable and evenly distributes the weight of the structure. It also protects the building from natural forces like wind, floods, and seismic activity.

7.3.1.2 Footing

Footings are a specific type of foundation element. They are typically wider at the base to spread the building load over a larger area, ensuring that the structure doesn't sink or settle unevenly. Footings are especially important in areas where soil conditions are weak or variable, as they help to prevent differential settlement, which can lead to structural damage.

7.3.2 Types of Foundations

The type of foundation used depends on the building type, the load it will carry, and the ground conditions.

- **Shallow Foundations (Spread Footings)** – These are used when the load-bearing capacity of the surface soil is sufficient to support the structure. Shallow foundations are common for smaller buildings like houses, offices, and light commercial structures.

Example – A residential house with a basement typically has a shallow foundation, consisting of strip or spread footings under the load-bearing walls. The footings distribute the load over a larger area of soil, preventing settlement and damage.

- **Deep Foundations (Piles or Caissons)** – When the surface soil lacks the strength to support a structure, deep foundations are used to transfer the load to deeper, more stable soil or rock layers. These are commonly used for large buildings, bridges, and structures on weak or compressible soils.

Example – A high-rise building in a city with soft or unstable soil will often use pile foundations. The piles are driven deep into the ground until they reach a stable layer, providing a solid base to support the enormous weight of the structure.

7.3.3 Types of Footings

There are different types of footings used based on the load and soil conditions:

- **Strip Footing** – Used for supporting load-bearing walls, this type of footing runs continuously along the length of the wall, spreading the load over a larger area.

Example – In the construction of a house, strip footings are used to support the walls, ensuring that the load from the roof, walls, and floors is evenly distributed.

- **Isolated or Pad Footing** – These are used to support individual columns or pillars. They are commonly square or rectangular and designed to carry the load of a single point load or column.

Example – In office buildings, isolated footings are often used under each column to distribute the load from the upper floors and roof.

- **Combined Footing** – When two or more columns are close to each other, a combined footing is used to support them. It distributes the load from both columns over a wider area.

Example – In a commercial building where columns are placed close together, combined footings may be used to provide support to multiple columns without constructing separate footings.

- **Raft or Mat Foundation** – Used when the soil is weak and the load from the building must be spread across the entire footprint of the structure, raft foundations are large concrete slabs that cover the whole building area.

Example – Hospitals and industrial buildings with large areas often use raft foundations because they provide a uniform load distribution across the entire structure.

- **Pile Footing** – These are used in areas where the topsoil is weak, and deep foundations are required. Pile footings transfer the load of the structure deep into the ground.

Example – A bridge constructed over a river would use pile footings driven into the riverbed to provide the necessary stability against shifting and flowing water.

7.3.4 Soil Considerations

The stability of any foundation depends heavily on the type of soil in which it is placed. Soil investigations are conducted to determine the load-bearing capacity, stability, and behavior of the ground. Different soils react differently under load—clay soils may expand and contract with moisture, while sandy soils may shift under pressure. Therefore, selecting the right foundation and footing depends on understanding the soil's properties.

Example – In a location with expansive clay soils, where the soil swells when wet and shrinks when dry, deep foundations such as piles are often preferred over shallow foundations to ensure stability.

7.3.5 Foundation Depth

The depth of the foundation is determined by the load it needs to carry, the type of soil, and the climate. Shallow foundations are typically used in stable, hard soils, while deeper foundations are required in areas with weak soils or where frost and temperature changes may cause ground movement.

Example – In cold climates where frost can penetrate deep into the ground, foundations are designed to be placed below the frost line to avoid frost heaving, which can damage the foundation and the structure it supports.

7.3.6 Water Table and Drainage

The water table level is a crucial factor in foundation design. If the water table is high, it can create hydrostatic pressure on the foundation, leading to water seepage and weakening of the structure. Proper drainage systems, such as French drains or sump pumps, may be needed to prevent water from affecting the foundation.

Example – A building located near a river may face challenges due to a high water table, and a deep foundation with waterproofing measures might be necessary to prevent water infiltration.

7.3.7 Foundation Materials

The materials used for foundations and footings vary based on the type of construction and the ground conditions. Common materials include reinforced concrete, masonry, and in some cases, treated timber or steel (for specific types of piles). Reinforced concrete is the most widely used material because of its strength, durability, and versatility.

Example – For a large industrial warehouse, reinforced concrete footings and foundations are ideal because they can support heavy machinery and withstand significant loads over time.

7.3.8 Summary

In summary, foundations and footings are critical elements in the design and construction of any structure. The choice of foundation type—whether shallow or deep—depends on factors like soil conditions, load requirements, water table level, and environmental conditions. Careful design and proper execution ensure that the building remains stable, safe, and functional for its intended lifespan. Examples from residential, commercial, and industrial settings illustrate how foundations and footings are tailored to meet different needs based on the specific conditions of each project.

7.4 TYPES OF BUILDINGS IN PLANT MAINTENANCE AND WORKS SERVICES

In the context of plant maintenance and works services, various types of buildings play a crucial role in ensuring that industrial processes run smoothly and efficiently. These buildings house different activities related to production, storage, administrative functions, and support services that are essential for the smooth operation of a plant. Each type of building requires specific maintenance approaches and strategies to ensure operational efficiency, safety, and longevity.

7.4.1 Production Buildings

These are the core structures in industrial plants where manufacturing, processing, or assembly takes place. Production buildings often contain heavy machinery, production lines, and equipment essential to the manufacturing process.

Example – In an automobile manufacturing plant, the production building houses conveyor belts, robotic arms, welding stations, and other machinery used to assemble cars.

Maintenance Focus – Regular inspection of machinery, cleaning, lubricating mechanical components, maintaining the electrical systems, and ensuring that ventilation and temperature control systems are functioning properly to prevent equipment failure and production delays.

7.4.2 Storage and Warehouse Buildings

These buildings are designed for storing raw materials, spare parts, and finished goods. Warehouses must be maintained to prevent damage to stored goods, ensure safety, and provide an efficient workflow for storing and retrieving items.

Example – A logistics and distribution center uses large warehouse buildings with pallet racking systems to store goods that will be distributed to various retailers.

Maintenance Focus – Checking the condition of storage racks, maintaining climate control systems (temperature and humidity), ensuring proper lighting, fire protection, and regularly inspecting forklifts or other material-handling equipment.

7.4.3 Administrative Buildings

These are office buildings where administrative, managerial, and planning activities take place. Administrative buildings need to be maintained to provide a safe, comfortable, and functional working environment for staff.

Example – In a large chemical plant, administrative buildings may include offices for engineers, accountants, and human resources personnel, who coordinate production schedules, manage staff, and handle procurement and logistics.

Maintenance Focus – Routine cleaning, HVAC system maintenance, lighting and electrical maintenance, plumbing services, and ensuring that safety systems such as fire alarms and sprinklers are operational.

7.4.4 Utility Buildings

Utility buildings house essential infrastructure for the entire plant, such as power generation equipment, boilers, chillers, water treatment systems, and backup generators. These buildings are critical because they support the utilities needed for the plant's production activities.

Example – In a power plant, utility buildings may house turbines, generators, transformers, and boilers responsible for generating electricity.

Maintenance Focus – Frequent inspection and maintenance of mechanical and electrical systems, checking power distribution systems, ensuring the proper functioning of boilers and chillers, and ensuring compliance with environmental and safety regulations.

7.4.5 Laboratory Buildings

Many industrial plants, especially those involved in pharmaceuticals, chemicals, and food processing, require laboratory buildings for quality control, research, and product development.

Example – A pharmaceutical plant might have laboratories for testing the quality of raw materials and finished products, as well as research labs for developing new drugs.

Maintenance Focus – Maintaining clean and sterile environments, ensuring proper ventilation and temperature control, calibrating laboratory equipment, checking safety systems like fume hoods and emergency showers, and regularly updating safety protocols.

7.4.6 Maintenance Workshops

These buildings are dedicated to repairing, maintaining, and fabricating parts or equipment used in the plant. The workshop houses tools, machinery, and equipment necessary for conducting maintenance tasks.

Example – In a steel manufacturing plant, the maintenance workshop may have welding equipment, machine tools, and spare parts required for repairing production machinery.

Maintenance Focus – Keeping tools and machines in working order, maintaining a safe and organized workspace, ensuring all safety equipment is functional, and keeping inventory of spare parts and consumables.

7.4.7 Training and Educational Facilities

Many industrial plants have buildings dedicated to employee training and development. These facilities are equipped with classrooms, workshops, and equipment for hands-on training.

Example – A large oil refinery may have training facilities where employees receive instruction on safety protocols, handling hazardous materials, and operating complex machinery.

Maintenance Focus – Regular upkeep of educational equipment, maintaining audiovisual and computer systems, ensuring that safety measures are in place, and maintaining the physical condition of classrooms and training areas.

7.4.8 Safety and Health Buildings

These buildings house safety and health equipment, emergency response systems, and medical facilities for plant employees. They are essential for responding to accidents, managing worker health, and complying with occupational health regulations.

Example – A coal mining facility might have safety buildings equipped with fire suppression equipment, first aid rooms, emergency showers, and eye wash stations.

Maintenance Focus – Ensuring that all safety equipment is operational and up to date, conducting regular safety drills, checking the availability and functionality of first aid equipment, and ensuring compliance with occupational health and safety laws.

7.4.9 Specialized Industrial Buildings

These are buildings designed for specific tasks or purposes within the plant, such as chemical storage, hazardous material handling, or assembly lines for specific products.

Example – A chemical plant may have specialized buildings for storing hazardous chemicals that require specific safety protocols, ventilation, and temperature control.

Maintenance Focus – Ensuring that safety systems such as fire suppression, ventilation, and chemical containment systems are functioning properly, regular inspections to prevent leaks or contamination, and compliance with environmental and safety regulations.

7.4.10 Summary

Different types of buildings within a plant serve various purposes, ranging from production to storage, administration, and safety. Each building requires a specific maintenance approach based on its function and the equipment housed within it. Proper maintenance ensures the reliability and efficiency of plant operations, reduces the risk of equipment failure, and enhances the safety of the working environment. Whether it's maintaining production machinery in manufacturing buildings or ensuring the proper functioning of safety systems in health buildings, regular upkeep and preventive maintenance are critical for the smooth running of industrial facilities.

7.5 ALIGNMENT IN PLANT MAINTENANCE AND WORKS SERVICES

Alignment refers to the precise adjustment of machinery, equipment, and components within an industrial plant to ensure they are correctly positioned relative to each other. In the context of plant maintenance and works services, alignment plays a crucial role in maintaining the efficient and safe operation of machinery and equipment, reducing wear and tear, enhancing performance, and preventing costly breakdowns. Poor alignment can lead to excessive vibration, increased energy consumption, and equipment failures, which can halt production and increase maintenance costs.

7.5.1 Importance of Alignment in Plant Maintenance

7.5.1.1 Increased Equipment Lifespan

Proper alignment of machinery helps to evenly distribute loads and forces, minimizing unnecessary stress on components such as shafts, bearings, and couplings. This can significantly extend the lifespan of the equipment, reducing the frequency of repairs and replacements.

7.5.1.2 Energy Efficiency

Misaligned machines consume more energy due to increased friction and resistance. Proper alignment ensures that machines operate with minimal energy loss, leading to improved energy efficiency and lower operating costs.

7.5.1.3 Reduced Downtime

Poor alignment can cause premature wear and mechanical failures, resulting in unexpected breakdowns and unplanned downtime. By maintaining proper alignment, plant maintenance teams can reduce the likelihood of equipment failure and keep production running smoothly.

7.5.1.4 Enhanced Safety

Misaligned equipment can generate excessive vibration, noise, and heat, which may create hazardous conditions in the workplace. Proper alignment reduces these risks, ensuring a safer environment for workers.

7.5.1.5 Improved Product Quality

In manufacturing processes, misaligned equipment can affect the precision and quality of the final product. For example, in processes such as metal cutting or printing, alignment issues can lead to dimensional inaccuracies, affecting the quality of the output. Maintaining proper alignment ensures that products are manufactured to the required specifications.

7.5.2 Types of Alignment in Plant Maintenance

7.5.2.1 Shaft Alignment

Shaft alignment involves aligning the rotational axis of two or more connected components, such as motors and pumps, to ensure that they operate on the same axis. Misaligned shafts can cause vibrations, increased wear on bearings, and coupling failures.

Example – In a centrifugal pump system, the alignment between the pump and the motor is critical. Misalignment can lead to excessive vibration and premature bearing failure. Laser alignment tools are often used to achieve precise shaft alignment.

7.5.2.2 Belt and Chain Alignment

Belt and chain drives transfer power between rotating equipment, such as motors and conveyor systems. Proper alignment of the pulleys or sprockets ensures smooth operation and prevents wear on the belts or chains.

Example – In a conveyor belt system used in material handling, misalignment of the belt can cause uneven wear, leading to belt damage and reduced efficiency. Regular alignment checks can prevent belt slippage and ensure the smooth movement of materials.

7.5.2.3 Bearing Alignment

Bearings support rotating parts, and their proper alignment is essential to reduce friction and wear. Misaligned bearings can cause uneven load distribution, leading to premature failure.

Example – In rotary kilns used in cement production, bearing alignment is crucial to support the massive rotating cylinder. Misalignment can cause uneven load distribution, leading to wear on the bearings and the kiln shell, affecting the overall operation.

7.5.2.4 Roller Alignment

In industries such as paper, textiles, and steel manufacturing, rollers are used to move or shape materials. Proper alignment of rollers ensures even pressure distribution, which is essential for achieving uniform product thickness or shape.

Example – In a paper mill, roller misalignment can cause uneven paper thickness or wrinkling. Regular alignment checks ensure that the rollers apply consistent pressure across the entire width of the paper.

7.5.3 Alignment Tools and Techniques

7.5.3.1 Laser Alignment

Laser alignment tools are widely used in modern plant maintenance to achieve precise alignment of machinery components. These tools project a laser beam along the axis of the equipment, allowing technicians to measure and correct any misalignment.

Example – Laser alignment is commonly used in aligning rotating equipment like motors and pumps, as it provides high accuracy and reduces the time required for alignment.

7.5.3.2 Dial Indicators

Dial indicators are mechanical devices that measure displacement or alignment errors by detecting slight movements or shifts in machine components.

Example – In machine tool maintenance, dial indicators are used to measure the runout of spindles, ensuring that they are properly aligned for precision machining.

7.5.3.3 Straightedge and Feeler Gauges

These traditional tools are used for checking the alignment of flat surfaces or components. A straightedge is placed across the components, and feeler gauges are used to check for gaps that indicate misalignment.

Example – In belt-driven equipment, straightedges can be used to align pulleys to ensure that the belts run smoothly without slipping or wearing out prematurely.

7.5.4 Examples of Alignment in Various Industries

7.5.4.1 Power Plants

In power plants, proper alignment of turbines, generators, and pumps is critical to ensure efficient energy production. Misalignment can lead to excessive vibration and increased wear on rotating parts, reducing the plant's overall efficiency.

7.5.4.2 Oil and Gas Industry

In refineries and drilling operations, proper alignment of pumps, compressors, and other rotating equipment is essential to prevent equipment failures and ensure safe and continuous operation. Misalignment in this industry can lead to catastrophic failures, resulting in significant downtime and potential safety hazards.

7.5.4.3 Manufacturing Industry

In industries such as automotive, aerospace, and heavy machinery, proper alignment of assembly line machinery, conveyor systems, and robotic arms is necessary to maintain product quality and ensure smooth production flow.

7.5.5 Conclusion

Alignment is a critical aspect of plant maintenance and works services, ensuring that machinery and equipment operate efficiently, safely, and with minimal wear. Regular alignment checks and adjustments are essential to prevent breakdowns, reduce maintenance costs, and extend the lifespan of equipment. Using modern alignment tools such as laser alignment systems, plant maintenance teams can achieve high levels of precision, leading to improved operational performance and reduced downtime. Proper alignment practices help maintain product quality, ensure safety, and contribute to the overall success of industrial operations.

7.6 INSTRUMENTATION IN PLANT MAINTENANCE AND WORKS SERVICES

Instrumentation plays a vital role in the efficient operation, monitoring, and maintenance of industrial plants. It involves the use of various devices and systems to measure, monitor, and control the operational variables of equipment and processes in a plant. These variables include pressure, temperature, flow rate, level, and others that are essential to ensure the smooth and safe functioning of machinery and industrial processes.

In the context of plant maintenance and works services, instrumentation helps maintain the plant's operational efficiency, ensures safety, enhances predictive maintenance, and aids in decision-making by providing real-time data and insights.

7.6.1 Importance of Instrumentation in Plant Maintenance

7.6.1.1 Monitoring and Control

Instrumentation provides continuous monitoring of critical parameters such as temperature, pressure, flow, and vibration, which are essential for the smooth functioning of equipment. Through sensors and control systems, it allows plant operators to monitor the performance of machinery in real-time, helping them to make adjustments as needed to prevent malfunctions or inefficiencies.

Example – In a steam boiler system, instrumentation such as pressure gauges, temperature sensors, and level indicators is used to monitor and control the steam production process, ensuring the system operates safely within the required parameters.

7.6.1.2 Preventive and Predictive Maintenance

Instrumentation enables the implementation of preventive and predictive maintenance strategies by providing data that can be analysed to predict potential equipment failures. This allows maintenance teams to take corrective actions before a breakdown occurs, reducing unplanned downtime and extending the life of equipment.

Example – Vibration sensors installed on rotating machinery like pumps and motors can detect early signs of imbalance or misalignment, prompting maintenance interventions before catastrophic failure occurs.

7.6.1.3 Improving Safety

Instrumentation is critical for ensuring safety in industrial plants. It monitors hazardous conditions, triggers alarms, and automatically shuts down equipment if unsafe conditions arise. This helps to protect both the machinery and personnel from accidents and injuries.

Example – In a chemical processing plant, gas detectors are used to monitor for the presence of dangerous gases like methane or hydrogen sulfide. If concentrations exceed safe levels, alarms are triggered, and automated shutdown systems may be activated to prevent accidents.

7.6.1.4 Efficiency and Optimization

Instrumentation systems contribute to operational efficiency by allowing precise control over industrial processes. By optimizing processes based on accurate real-time data, plants can reduce energy consumption, minimize waste, and increase productivity.

Example – In oil refining, flow meters, temperature controllers, and pressure gauges ensure that the refining process operates at optimal conditions, maximizing output while minimizing energy use.

7.6.1.5 Data Logging and Analysis

Instrumentation systems often include data logging features that record historical data on equipment performance and operational conditions. This data can be analysed to identify trends, optimize processes, and improve maintenance schedules.

Example – In power plants, data from temperature and pressure sensors is logged over time and analysed to optimize turbine performance and fuel efficiency.

7.6.2 Types of Instrumentation in Plant Maintenance

7.6.2.1 Temperature Sensors

These devices measure the temperature of equipment and processes. Common types include thermocouples, resistance temperature detectors (RTDs), and infrared sensors.

Example – In furnace operations, temperature sensors are used to monitor and regulate the internal temperature to ensure proper material processing without overheating or underheating.

7.6.2.2 Pressure Sensors

Pressure sensors measure the pressure of liquids or gases in a system. They are critical in processes involving fluids, gases, and hydraulic systems.

Example – In hydraulic systems, pressure sensors ensure that the hydraulic fluid is within the correct pressure range to avoid system failure.

7.6.2.3 Flow Meters

Flow meters measure the rate at which fluids or gases move through pipes and channels. Accurate flow measurement is essential for processes like water treatment, oil refining, and chemical manufacturing.

Example – In a water treatment plant, flow meters are used to measure the flow rate of water through the filtration system, ensuring proper treatment.

7.6.2.4 Level Sensors

Level sensors measure the level of materials, liquids, or solids, in tanks or silos. This is crucial for inventory management and process control.

Example – In chemical storage tanks, level sensors ensure that chemicals are stored safely and within the required capacity limits to prevent overflow or shortage.

7.6.2.5 Vibration Sensors

These sensors detect vibrations in machinery, often used in rotating equipment such as motors, turbines, and pumps. Excessive vibration can indicate issues like misalignment, imbalance, or bearing wear.

Example – In wind turbines, vibration sensors help monitor the condition of the gearbox and rotor to ensure smooth operation and prevent premature failure.

7.6.2.6 Proximity Sensors

Proximity sensors detect the presence or absence of objects or components in equipment. They are often used for safety purposes and in automated machinery.

Example – In conveyor belt systems, proximity sensors detect the presence of objects on the belt, ensuring that materials are correctly processed or sorted.

7.6.2.7 Control Systems

Control systems, such as programmable logic controllers (PLCs) and distributed control systems (DCS), are used to automate and manage industrial processes based on data received from various sensors. These systems enable efficient and accurate process control.

Example – In automated manufacturing lines, PLCs are used to control the sequence of operations based on sensor inputs, ensuring product consistency and quality.

7.6.3 Examples of Instrumentation in Various Industries

7.6.3.1 Oil and Gas Industry

In the oil and gas sector, instrumentation is essential for controlling drilling operations, refining processes, and pipeline monitoring. Pressure, temperature, and flow sensors are used extensively to ensure safe and efficient operation.

7.6.3.2 Pharmaceutical Industry

In pharmaceutical manufacturing, precision control of temperature, pressure, and flow is critical for maintaining the quality of products. Instruments like cleanroom particle counters, temperature sensors, and flow meters are used to meet strict regulatory requirements.

7.6.3.3 Power Generation

In power plants, instrumentation is used to monitor turbine performance, boiler operation, and electrical output. Pressure gauges, temperature sensors, and vibration monitoring systems help to optimize performance and prevent failures.

7.6.4 Summary

In summary, instrumentation in plant maintenance and works services is essential for ensuring the efficient operation, safety, and reliability of industrial processes. It provides real-time data, enabling operators to monitor and control critical parameters like temperature, pressure, flow, and vibration. Through the use of advanced sensors and control systems, instrumentation supports preventive and predictive maintenance strategies, improves energy efficiency, and enhances safety. By implementing proper instrumentation, plants can optimize their processes, reduce operational costs, and extend the lifespan of their equipment.

CHAPTER EIGHT

8.0 COMPREHENSIVE EQUIPMENT MAINTENANCE: LIFECYCLE MANAGEMENT, LUBRICATION, REPAIRS, AND SERVICING IN INDUSTRIAL OPERATIONS

8.1 WEAR AND SERVICE LIFE OF INDUSTRIAL EQUIPMENT AND MACHINERY

Wear refers to the gradual degradation of materials or components in industrial equipment and machinery due to continuous use, environmental factors, or operational conditions. Over time, wear affects the efficiency, reliability, and overall performance of machines, leading to increased maintenance requirements, downtime, and eventually, the end of the service life of the equipment. The service life of industrial machinery refers to the expected duration or operational lifespan during which the equipment performs its intended function effectively and safely before it needs to be repaired, refurbished, or replaced.

Understanding the causes and types of wear, as well as the factors that influence the service life of machinery, is critical in plant maintenance, as it helps in implementing preventive maintenance strategies and extending the operational lifespan of the equipment.

8.1.1 Types of Wear in Industrial Equipment

8.1.1.1 Abrasive Wear

Abrasive wear occurs when hard particles or rough surfaces come into contact with the equipment's moving parts, causing material to be scraped or worn away. This type of wear is common in industries dealing with materials like sand, gravel, and minerals.

Example – In mining equipment such as crushers and conveyors, abrasive wear is prevalent due to the constant contact with rocks and minerals, resulting in erosion of the metal components.

8.1.1.2 Adhesive Wear

Adhesive wear happens when two surfaces slide over each other under pressure, causing material from one surface to transfer to the other. This can lead to surface damage, roughness, and even welding of the surfaces under extreme conditions.

Example – Bearings and shafts in industrial machines may experience adhesive wear when there is insufficient lubrication, leading to metal-on-metal contact.

8.1.1.3 Fatigue Wear

Fatigue wear results from cyclic loading and unloading of components, leading to the formation of cracks and eventual failure. Repeated stresses cause microscopic cracks to form, which grow over time until the component breaks.

Example – In rotating machinery like turbines and gear systems, fatigue wear is common due to the constant cyclical forces acting on the components.

8.1.1.4 Corrosive Wear

Corrosive wear is caused by chemical reactions between the equipment's surface and the surrounding environment, leading to material degradation. Corrosion can weaken metal components and shorten the lifespan of the equipment.

Example – In chemical processing plants, pipelines and vessels are prone to corrosive wear due to exposure to acidic or caustic substances.

8.1.1.5 Erosive Wear

Erosive wear occurs when high-velocity fluids or particles strike the surface of a material, leading to material loss. This is common in industries where machinery handles gases or liquids at high velocities.

Example – In pumps and fan blades, erosive wear can happen due to the impact of high-speed particles or fluid flow, especially in industries dealing with liquids or gas handling.

8.1.2 Factors Affecting the Service Life of Industrial Equipment

8.1.2.1 Material Properties

The type of material used to manufacture equipment directly impacts its resistance to wear and its overall durability. For example, equipment made from hardened steel or ceramic-coated parts may have a longer service life compared to softer metals, especially in abrasive environments.

8.1.2.2 Operating Conditions

Harsh operational environments, such as extreme temperatures, humidity, and exposure to chemicals, can accelerate wear and reduce the service life of machinery. Equipment that operates under extreme pressure or high-speed conditions is more likely to experience rapid wear.

Example – In steel mills, machinery is subjected to high temperatures and continuous operation, which can accelerate wear and tear on critical components.

8.1.2.3 Maintenance Practices

Proper maintenance is essential to extend the service life of industrial machinery. Regular lubrication, cleaning, and adjustment of equipment help prevent excessive wear and avoid premature failure.

Example – A conveyor system with regular lubrication and belt tension adjustments will experience less wear and a longer service life compared to one that is poorly maintained.

8.1.2.4 Load and Stress

Operating machinery beyond its designed load or stress capacity can significantly shorten its service life. Continuous overloading can lead to fatigue wear, overheating, and premature failure.

Example – In construction equipment like bulldozers or cranes, operating under heavy loads continuously can cause fatigue wear in the hydraulic systems and structural components.

8.1.2.5 Lubrication

Lubrication reduces friction between moving parts and minimizes adhesive and abrasive wear. Insufficient or improper lubrication can lead to excessive wear and reduce the operational lifespan of machinery.

Example – Gearboxes in heavy machinery that are properly lubricated experience significantly less wear than those operating with contaminated or insufficient lubricants.

8.1.2.6 Vibration and Misalignment

Excessive vibration or misalignment of components in machinery can lead to accelerated wear. It can cause bearings, shafts, and other rotating parts to wear unevenly or fail prematurely.

Example – In rotating machinery like motors or turbines, even slight misalignments can lead to imbalanced forces and vibrations that accelerate wear in bearings and couplings.

8.1.3 Strategies to Extend the Service Life of Equipment

8.1.3.1 Preventive Maintenance

Regular inspections, maintenance tasks such as lubrication, cleaning, and adjustments help identify wear issues before they result in failure. Preventive maintenance aims to prolong the service life of equipment by addressing minor issues early.

Example – Scheduled maintenance of HVAC systems in a manufacturing plant includes replacing filters, cleaning coils, and lubricating fans, which prolongs the life of the system.

8.1.3.2 Predictive Maintenance

Using technologies such as vibration analysis, thermal imaging, and oil analysis, predictive maintenance aims to detect early signs of wear or potential failure, allowing for timely intervention.

Example – Vibration monitoring of rotating pumps in a refinery can detect imbalances or bearing wear before a major breakdown occurs, enabling predictive maintenance.

8.1.3.3 Proper Training and Operation

Ensuring that machinery is operated within its designed parameters and that operators are properly trained can prevent excessive wear and misuse, which can extend the service life of the equipment.

Example – Training operators to avoid overloading forklifts can prevent mechanical strain and wear on the hydraulic systems and structural components.

8.1.3.4 Upgrading Materials or Components

Replacing worn-out components with more durable materials or upgrading to advanced technology can extend the lifespan of equipment. For example, replacing traditional bearings with ceramic bearings can improve wear resistance and reduce maintenance costs.

8.1.4 Summary

Wear and service life of industrial equipment and machinery are crucial factors that determine the efficiency, reliability, and cost-effectiveness of industrial operations. Understanding the types of wear and factors that influence the lifespan of machinery allows for better maintenance practices, such as preventive and predictive maintenance, which can significantly reduce downtime and extend equipment life. By carefully monitoring wear and ensuring proper maintenance, industries can enhance productivity and reduce the total cost of ownership for their equipment.

8.2 PLANNED LUBRICATION: TYPES AND EFFECTS

Planned lubrication is a systematic approach to ensuring that machinery and equipment receive the correct type and amount of lubricant at appropriate intervals. It is a key component of preventive maintenance that helps minimize friction, reduce wear, and extend the service life of industrial equipment. A well-planned lubrication schedule can prevent unplanned downtime, reduce maintenance costs, and improve equipment performance.

Lubrication is critical because it reduces the friction between moving parts, which, if unchecked, can lead to excessive wear, heat generation, and component failure. The correct selection and application of lubricants are crucial to achieving these objectives.

8.2.1 Types of Lubrication

8.2.1.1 Grease Lubrication

Definition – Grease is a semi-solid lubricant made from oil and a thickening agent, often used where liquid lubricants are difficult to apply or tend to run off.

Application – Grease is

commonly used in bearings, joints, gears, and areas exposed to contaminants like dust and moisture. It stays in place longer than oil, making it ideal for components that are not easily accessible.

Effects – Grease forms a thicker film that helps protect surfaces from dirt and contaminants. It also lasts longer than oil in certain applications, which can reduce the frequency of lubrication.

Example – In wheel bearings of industrial trucks, grease lubrication is preferred due to its ability to stay in place under high load conditions and to resist the ingress of contaminants such as dirt or water.

8.2.1.2 Oil Lubrication

Definition – Oil is a liquid lubricant that reduces friction between moving parts and helps dissipate heat generated from friction.

Application – Oil is used in engines, compressors, gearboxes, and hydraulic systems, where it can flow easily into tight spaces and continuously lubricate moving parts.

Effects – Oil lubrication reduces friction efficiently and helps to cool moving parts by absorbing and dissipating heat. It also aids in flushing out contaminants from the lubrication points.

Example – Hydraulic systems in industrial machinery rely on oil lubrication to ensure smooth operation of the system components and reduce wear in pumps, valves, and cylinders.

8.2.1.3 Dry Lubrication

Definition – Dry lubricants, such as graphite or molybdenum disulfide, are used where traditional oil or grease lubricants may attract dust or contaminants.

Application – Dry lubricants are often applied in high-temperature environments or where cleanliness is critical, such as in food processing or textile manufacturing.

Effects – Dry lubricants reduce friction without attracting dust and debris, making them suitable for applications where cleanliness is essential or where extreme temperatures render traditional lubricants ineffective.

Example – Conveyor belts in a food processing plant may use dry lubricants to prevent contamination from oils or greases while maintaining smooth operation of moving parts.

8.2.1.4 Boundary Lubrication

Definition – Boundary lubrication occurs when a thin film of lubricant separates two surfaces under heavy load conditions, preventing direct metal-to-metal contact.

Application – This type of lubrication is common in gearboxes, cams, and bearings where extreme pressure is exerted on surfaces.

Effects – Boundary lubrication helps protect surfaces from wear and minimizes friction when the lubricant film is thin due to high pressure.

Example – Gears in heavy-duty industrial machinery experience boundary lubrication when operating under high load conditions, where the thin film of oil is still able to protect the surfaces from direct contact.

8.2.1.5 Hydrodynamic Lubrication

Definition – Hydrodynamic lubrication occurs when a thick film of lubricant completely separates the moving parts, creating a cushion that prevents contact.

Application – This is typical in rotating machinery such as turbines, pumps, and large motors, where the lubricant film is maintained by the relative motion of the surfaces.

Effects – Hydrodynamic lubrication virtually eliminates metal-to-metal contact, reducing friction and wear. It also aids in cooling components by carrying away heat.

Example – In large hydroelectric turbines, hydrodynamic lubrication ensures that the rotating shafts and bearings are fully separated by a thick oil film, which reduces wear and increases efficiency.

8.2.2 Effects of Planned Lubrication

8.2.2.1 Reduced Wear and Friction

Properly planned lubrication minimizes direct contact between metal surfaces, reducing friction and wear. This leads to longer equipment life and reduces the likelihood of breakdowns.

Example – Regular oiling of industrial chains prevents the metal links from rubbing directly against one another, which prolongs their service life and maintains operational efficiency.

8.2.2.2 Improved Heat Dissipation

Lubricants carry away heat generated by friction, preventing overheating and ensuring that machinery operates within safe temperature ranges.

Example – In high-speed compressors, oil is used not only to lubricate the moving parts but also to remove the heat generated by the high-speed rotation.

8.2.2.3 Protection from Contaminants

Lubricants, especially grease, can form a protective barrier that prevents contaminants like dust, dirt, and water from entering sensitive parts of machinery.

Example – Outdoor machinery, such as construction cranes, benefit from grease lubrication in their joints to prevent dirt and water from entering and damaging the moving parts.

8.2.2.4 Extended Equipment Life

Regular and planned lubrication significantly extends the lifespan of equipment by preventing premature wear and failure of components. It also reduces the frequency of repairs and replacements.

Example – In automotive assembly lines, planned lubrication of robotic arms ensures consistent performance over time, reducing downtime for maintenance.

8.2.2.5 Reduced Energy Consumption

Proper lubrication reduces the frictional forces between moving parts, which, in turn, reduces the amount of energy required to operate the machinery. This can lead to significant cost savings, especially in large-scale industrial operations.

Example – A well-lubricated conveyor system in a manufacturing plant requires less energy to move materials compared to one where friction from poorly maintained rollers increases energy consumption.

8.2.2.6 Minimized Downtime

Planned lubrication helps avoid unexpected machinery failures, leading to fewer disruptions in production. Preventive lubrication schedules keep machines running smoothly, allowing for planned downtime rather than emergency repairs.

Example – In the textile industry, planned lubrication of looms and spinning machines ensures continuous operation, preventing sudden breakdowns that could halt production.

8.2.3 Summary

In summary, planned lubrication is essential for maintaining the efficiency, performance, and longevity of industrial machinery and equipment. By using the appropriate type of lubricant and applying it at the right intervals, plant operators can reduce friction, wear, and overheating, thereby improving the overall productivity of the plant. Regular lubrication also protects machinery from contaminants, extends equipment life, and reduces operational costs. In industries where uptime is critical, such as manufacturing and energy production, implementing a planned lubrication program is key to sustaining operations and achieving long-term success.

8.3 ASSEMBLY AND DISASSEMBLY OF MECHANISMS AND MACHINES

The assembly and disassembly of mechanisms and machines involve the process of putting together or taking apart various components that make up mechanical systems. These procedures are essential in maintenance, repair, and upgrading equipment. Assembly and disassembly are particularly important for ensuring that machines function properly, are aligned correctly, and meet safety and operational standards.

8.3.1 Key Steps in Assembly and Disassembly

1 Preparation and Safety Precautions

- Prior to beginning, it's essential to understand the machinery, consult operation manuals, and ensure safety procedures.
- Proper tools and safety equipment (like gloves, safety goggles, and anti-static mats) should be selected and arranged for efficient workflow.

2 Documentation and Labeling

- Proper documentation, including photos or diagrams, is essential to track each part's location and orientation.
- Labels or tags help identify parts, especially in complex machinery, for accurate reassembly.

3 Disassembly Process

- Begin by isolating the power source and disassembling parts in sequence, typically starting from outer or accessible parts to more internal components.
- The process involves carefully unbolting, unfastening, or removing each part without causing damage.

4 Inspection and Cleaning

- During disassembly, each part is inspected for signs of wear, corrosion, or damage.
- Cleaning and lubricating parts as they are disassembled can prevent future breakdowns.

5 Reassembly Process

- During reassembly, parts are put back in the reverse order of disassembly, following the documentation carefully.
- All bolts, screws, and parts are tightened to specification, and alignment is verified for each stage.

6 Testing and Calibration

After reassembly, the machine is tested for functionality to confirm proper assembly. Calibration may be required to ensure it operates according to original specifications.

8.3.2 Examples of Assembly and Disassembly

8.3.2.1 Automobile Engines

Disassembly – This process typically involves removing external components like belts, hoses, and alternators before reaching the internal components, such as pistons, valves, and crankshaft.

Reassembly – Each part must be cleaned, inspected, and reinstalled precisely, following manufacturer specifications to ensure the engine runs efficiently.

8.3.2.2 Conveyor Belts in Manufacturing

Disassembly – Conveyor belts may need to be disassembled for maintenance or repairs, starting with motor detachment, followed by belt removal and inspection of rollers and gears.

Reassembly – During reassembly, components must be realigned to ensure smooth operation, as misalignment can cause wear and lead to frequent repairs.

8.3.2.3 Hydraulic Systems in Heavy Equipment

Disassembly – Hydraulic systems require special care during disassembly. Pressure should be relieved, and each line disconnected carefully to prevent leaks and contamination.

Reassembly – The reassembly process involves reconnecting hydraulic lines with secure fittings and precise alignment. The system is then bled to remove air pockets before testing.

8.3.2.4 Electric Motors

Disassembly – Electric motors require careful handling during disassembly. After disconnecting the power source, the outer casing is removed, followed by stator and rotor components, which are checked for wear.

Reassembly – Aligning the stator and rotor precisely is crucial for motor efficiency. Parts must be cleaned, and insulation checked, especially in high-heat motors.

8.3.3 Benefits of Effective Assembly and Disassembly

- **Improved Equipment Reliability**

Regular disassembly and inspection prevent premature breakdowns and increase equipment reliability, as in automobile engine maintenance.

- **Prolonged Machine Life**

Proper assembly practices reduce wear and prevent misalignment, prolonging equipment life. For example, reassembling conveyor belts correctly minimizes wear on the belt and rollers.

- **Safety and Compliance**

Safe and proper assembly helps comply with industry standards and minimizes risks of accidents, particularly in hydraulic or heavy machinery applications.

8.3.4 Summary

In summary, the assembly and disassembly of mechanisms and machines are fundamental maintenance activities that support operational efficiency, safety, and equipment longevity. Proper documentation, careful inspection, and adherence to procedures allow machines to operate efficiently and reduce the frequency and cost of repairs. Through regular and accurate assembly and disassembly practices, industries ensure equipment reliability, optimize performance, and maintain safety standards, all of which are crucial for sustainable plant operations.

8.4 EQUIPMENT REPAIR AND ACCEPTANCE PROCEDURES

Equipment repair and acceptance procedures are systematic steps followed in industrial settings to ensure that repairs meet required standards, functionality, and safety. These procedures help verify that repaired equipment functions as intended and that repairs are completed correctly before the equipment is returned to operation. Following standard repair and acceptance procedures enhances equipment reliability, reduces downtime, and ensures safe operating conditions.

8.4.1 Key Phases of Equipment Repair

- **Initial Inspection and Diagnosis**

Equipment repair begins with an assessment of the faulty equipment to identify the root cause of the issue.

This includes visual inspections, diagnostic tests, and condition analysis.

Example – In diagnosing a malfunctioning pump, a technician might inspect for visible wear, check for fluid leaks, and use vibration analysis to detect internal bearing issues.

- **Repair Planning and Approval**

Once diagnosed, a repair plan is formulated, detailing steps, required tools, replacement parts, and estimated repair time.

Approval from relevant supervisors or engineers ensures the repair plan aligns with safety and budget requirements.

Example – For a damaged conveyor belt, the repair plan may outline part replacements and re-alignment steps, along with an estimated repair time to minimize production delays.

- **Execution of Repair**

The repair team then follows the approved repair plan. This can involve replacing faulty parts, performing corrective maintenance, cleaning, and adjusting components to manufacturer specifications.

Example – For an industrial motor, the repair may involve replacing worn bearings, rewinding coils, and re-aligning the rotor.

- **Post-Repair Inspection and Testing**

After the repair, thorough inspection and testing validate that all components function correctly and meet safety and performance standards.

Tests may include functionality tests, load tests, vibration analysis, and temperature monitoring.

Example – In a boiler system repair, post-repair tests might involve pressure testing to confirm no leaks and temperature monitoring to ensure it meets operational standards.

- **Documentation of Repairs**

Detailed documentation of the repair process, parts used, test results, and any adjustments made is crucial.

This serves as a reference for future maintenance and ensures accountability in case of future breakdowns.

Example – In repairing a hydraulic press, documentation may include part serial numbers, torque settings, fluid types, and inspection notes.

- **Acceptance and Handover**

The final phase involves a formal acceptance process, where the maintenance team presents the repaired equipment to the relevant supervisors or users.

Acceptance testing by operators or supervisors ensures satisfaction with the repair quality and verifies the equipment meets operational requirements.

Example – A repaired forklift is test-driven by the operator under supervision to confirm stability, manoeuvrability, and functionality.

8.4.2 Acceptance Criteria and Procedures

- **Functionality Tests**

Repaired equipment must operate as expected under standard load conditions.

Example – After repairing a CNC machine, sample parts are machined to ensure dimensional accuracy and repeatability.

- **Safety Compliance Checks**

Safety features such as emergency stops, guards, and limit switches must be inspected and tested to confirm they are in working order.

Example – For a repaired industrial lathe, emergency stops and safety guards are checked to ensure compliance with safety protocols.

- **Performance Testing**

Equipment may undergo performance testing under maximum operating conditions to ensure repairs hold up under stress.

Example – After repairing an air compressor, it is run at full capacity for a specific time to confirm it can maintain pressure and avoid overheating.

- **Operator Approval**

Once tests are passed, operators perform a trial run to verify they can operate the equipment as before.

Example – Following repair on an assembly line robotic arm, operators run a few cycles to confirm alignment with production requirements.

- **Sign-Off and Documentation**

Once all parties are satisfied, an official sign-off completes the acceptance process, with records updated to reflect the repair.

Example – For a repaired HVAC system, facility managers document the repair details and sign-off, then update maintenance logs and schedules.

8.4.3 Examples of Equipment Repair and Acceptance in Various Industries

- **Manufacturing Industry**

In manufacturing, frequent repairs and acceptance procedures ensure machines like CNC mills, drills, and robotic arms perform at optimal capacity.

Example – After repairing a robotic arm, a technician calibrates the arm's movements and conducts trial runs with sample workpieces. Operators verify it meets precision standards before signing off.

- **Construction Industry**

Construction equipment such as cranes, bulldozers, and excavators are critical to project timelines, making efficient repair and acceptance essential.

Example: After repairing a hydraulic leak in an excavator, hydraulic pressure and movement are tested. The operator then completes a digging test to confirm functionality before resuming use.

- **Energy and Utilities Industry**

Equipment such as turbines, generators, and transformers undergo rigorous testing after repairs to avoid potential downtime in energy supply.

Example – After rewinding a generator, technicians perform voltage and current tests, followed by a short operational run to confirm it meets required capacity.

8.4.4 Importance of Equipment Repair and Acceptance Procedures

Reliability – Ensures equipment functions as intended, reducing unexpected downtime and optimizing productivity.

Safety – Verifies repairs uphold safety standards, protecting operators and other personnel.

Cost-Efficiency – Reduces the likelihood of repeat repairs, which minimizes costs and maximizes equipment lifespan.

Accountability – Proper documentation provides a clear record of repairs, helping teams trace issues if future breakdowns occur.

8.4.5 Summary

In summary, by following systematic repair and acceptance procedures, industries ensure that repaired equipment performs reliably, maintains safety standards, and aligns with operational requirements.

8.5 MAINTENANCE SERVICING, AND UPKEEP OF EQUIPMENT IN INDUSTRIAL SETTINGS

Maintenance, servicing, and upkeep of equipment in industrial settings are crucial to ensuring operational efficiency, safety, and longevity. Each type of machinery—whether steam boilers, machine tools, turbines, pumps, conveyors, cranes, compressors, or valves—requires specific maintenance tasks suited to its function, wear, and operating conditions.

8.5.1 Steam Boilers

Maintenance Tasks – Steam boilers require periodic inspection and cleaning of boiler tubes, combustion checks, water quality testing, and safety valve testing.

Servicing – Regular descaling to prevent scale build-up, which affects heat transfer, and replacement of faulty gaskets or seals are essential. Annual shutdown inspections are typically done to clean internal surfaces and inspect for corrosion or leaks.

Upkeep – Monitoring water levels and maintaining correct boiler pressure help ensure the boiler operates efficiently and safely. Proper treatment of feed water is necessary to prevent corrosion and scaling.

Example – In a power plant, steam boilers are inspected quarterly, with detailed checks for scaling and water alkalinity, ensuring optimal steam quality for power generation.

8.5.2 Machine Tools

Maintenance Tasks – Regular lubrication of moving parts, checking tool alignment, and inspecting cutting tools for wear are essential to maintain accuracy and prevent damage.

Servicing – Parts like bearings and belts should be checked and replaced as needed to maintain precision and reliability. The calibration of machine tools is periodically verified to prevent deviations in production quality.

Upkeep – Daily cleaning and removal of debris and metal shavings help avoid damage and ensure accuracy in machining.

Example – In a CNC machine tool, monthly calibration and weekly lubrication checks are essential for maintaining precision and extending tool life.

8.5.3 Turbines

Maintenance Tasks – Turbines require regular inspection of blades, lubrication of moving parts, and alignment checks to avoid vibration-related damage.

Servicing – Periodic replacement of seals, bearings, and lubrication systems is necessary to prevent leaks and maintain efficiency.

Upkeep – Monitoring operational parameters, such as pressure and temperature, and conducting vibration analysis help in early detection of issues like blade erosion or misalignment.

Example – In a wind farm, turbines undergo biannual inspections to check blade conditions and gearbox alignment, as this ensures consistent power output.

8.5.4 Pumps

Maintenance Tasks – Routine lubrication, checking for leaks, and inspecting seals and couplings are critical to ensure pumps operate smoothly.

Servicing – Components such as seals and impellers may need periodic replacement to maintain optimal performance, especially in high-pressure or high-volume systems.

Upkeep – Monitoring flow rates and pressure levels can help detect issues early, such as cavitation or abnormal vibrations, which can lead to pump failure.

Example – In a water treatment plant, centrifugal pumps are inspected weekly for leaks and noise, with seals replaced annually to ensure continuous operation.

8.5.5 Conveyors

Maintenance Tasks – Routine checks for wear and tear on belts, chains, and rollers help in detecting potential failures. Lubrication of moving parts is essential for smooth operation.

Servicing – Bearings, rollers, and motor drive systems require servicing to avoid unexpected downtime. Adjusting tension on belts and chains is also critical to maintain efficiency.

Upkeep – Regular cleaning prevents debris buildup that can cause jams and reduce operational life.

Example – In a mining operation, conveyor belts are cleaned and inspected weekly for alignment and tension, with bearings lubricated monthly to handle the load of transported ore.

8.5.6 Cranes

Maintenance Tasks – Routine inspection of hoisting mechanisms, checking wire ropes, chains, and safety latches, and lubricating the crane's components are essential to safe crane operation.

Servicing – Hydraulic systems and load brakes should be serviced and replaced as needed to prevent accidents. Control systems and safety features require testing and adjustment to ensure reliability.

Upkeep – Daily inspections for visible signs of wear and weekly testing of emergency stop functions are common practices to maintain safety.

Example – In a construction yard, cranes are inspected daily for load chain condition, hydraulic levels, and emergency stop functionality to ensure safe operation.

8.5.7 Compressors

Maintenance Tasks – Compressors need regular lubrication, inspection of air filters, and checks on pressure settings and safety valves.

Servicing – Periodic replacement of filters, worn-out seals, and checking for air leaks is crucial to maintaining efficiency and preventing contamination of compressed air.

Upkeep – Monitoring temperature and pressure during operation helps in early detection of potential failures, reducing downtime.

Example – In a manufacturing plant, air compressors are serviced monthly to clean or replace filters, maintain pressure levels, and inspect seals, ensuring reliable compressed air supply.

8.5.8 Valves

Maintenance Tasks – Valves require regular inspections for leaks, checking for corrosion, and testing of opening and closing mechanisms.

Servicing – Routine servicing may involve cleaning or replacing seals, greasing valve stems, and ensuring that actuator mechanisms function properly.

Upkeep – Adjusting packing and replacing damaged parts are common practices to ensure valves maintain a leak-tight seal and control flow effectively.

Example – In a chemical processing plant, pressure relief valves are inspected quarterly for corrosion, with seals and seats replaced annually to ensure they operate safely under varying pressures.

8.5.9 Summary

In summary, these maintenance practices help optimize the efficiency, safety, and service life of each type of industrial equipment and machinery, leading to reduced downtime, enhanced productivity, and compliance with safety regulations. Regular upkeep, proper documentation, and adherence to established maintenance schedules are critical for smooth operations in industrial settings.

CHAPTER NINE

9.0 GENERAL OPERATIONS AND MAINTENANCE OF HIGH-SPEED ENGINES

Operating and maintaining high-speed engines requires careful attention to both general operation protocols and stringent safety inspections. High-speed engines are commonly used in industries such as automotive, power generation, marine, and manufacturing, where rapid and efficient power output is essential. However, their high operating speeds and intensive use make them prone to specific operational and safety concerns.

9.1 OPERATIONS, SAFETY INSPECTION, AND APPLICATIONS OF HIGH-SPEED MACHINES

9.1.1 General Operations of High-Speed Engines

1 Start-Up Procedures

Pre-Checks – Before starting a high-speed engine, operators should check fuel levels, oil pressure, and coolant levels. Confirming that all systems are in optimal condition helps avoid operational issues at high speeds.

Warm-Up – High-speed engines benefit from a gradual warm-up to operating temperature to ensure proper lubrication and prevent sudden load application, which can cause stress on engine components.

2 Monitoring Engine Parameters

Temperature and Pressure – Continuous monitoring of exhaust temperature, coolant temperature, and oil pressure is critical. High temperatures or pressure drops can indicate wear, potential overheating, or insufficient lubrication.

RPM Control – RPM (revolutions per minute) should be kept within manufacturer-recommended limits to avoid excessive wear or failure of components.

Fuel Efficiency – Monitoring fuel consumption is essential for high-speed engines. Abrupt increases in fuel usage could signal performance issues, including combustion inefficiencies or injector problems.

3 Load Management

High-speed engines should not be overloaded. Gradual load increase after the engine has warmed up is ideal, as rapid load changes can lead to imbalance and increased wear.

Lubrication – Proper lubrication is essential for minimizing friction and wear in high-speed engines. Operators should ensure adequate oil supply and monitor oil quality, as contaminated or degraded oil can damage engine components.

Shutdown Procedures – Shutting down a high-speed engine gradually, allowing it to cool, helps prevent thermal stress and protects vital engine parts. An abrupt shutdown may lead to component deformation due to heat.

9.1.2 Safety Inspection of High-Speed Engines

1 Scheduled Safety Inspections

Daily Inspections – These typically include checking oil levels, fuel levels, coolant levels, and general cleanliness around the engine.

Weekly Inspections – These may include testing and adjusting throttle controls, fuel injectors, and verifying the proper operation of safety systems such as automatic shutdown mechanisms.

Monthly or Periodic Inspections – At longer intervals, a comprehensive check of the engine should be conducted, including a detailed inspection of fuel injectors, valves, pistons, and compression ratios.

2 Specific Safety Checks

Overheating Prevention – High-speed engines can generate extreme heat, so cooling systems must be regularly inspected. This includes verifying radiator efficiency, coolant levels, and ensuring that no obstructions hinder airflow.

Vibration Analysis – High-speed engines can produce vibrations that, over time, may loosen components or cause structural damage. Regular vibration analysis helps detect potential misalignments or bearing wear.

Exhaust and Emissions Check – Emissions inspections are important for safety and regulatory compliance. Excessive smoke or abnormal colors in exhaust may indicate incomplete combustion or an internal fault.

Inspection of Safety Mechanisms – High-speed engines often feature emergency stop buttons, temperature sensors, and pressure-relief valves. These mechanisms should be regularly tested to confirm they operate correctly.

3 Wear and Component Inspection

Belts and Pulleys – Belts used in accessory systems should be inspected for signs of wear or slippage. Tension should be checked and adjusted if necessary.

Internal Components – Periodic teardown inspections of pistons, bearings, and connecting rods help identify wear and fatigue before they lead to failure.

Electrical and Ignition Systems – Faulty wiring or poor ignition timing can affect high-speed engines. Regular checks ensure that spark plugs, ignition coils, and wires are functioning properly.

9.1.3 High-Speed Engine Applications

High-speed engines are designed to operate at high revolutions per minute (RPM), typically above 1,000 RPM, and are widely used in applications where power, efficiency, and quick response are critical. Some examples of high-speed engine applications across various industries are outlined as follows:

1 Automotive Industry

Sports Cars – High-performance sports cars use high-speed engines to achieve fast acceleration and top speeds. For example, the engines in Ferrari, Lamborghini, and Porsche vehicles can rev up to 9,000 RPM or higher, delivering high power output.

Motorcycles – High-speed engines are common in sport and racing motorcycles. Brands like Yamaha, Ducati, and Kawasaki use high-speed engines to offer quick throttle response and top-notch performance, often exceeding 12,000 RPM.

Trucks and Buses – Many commercial vehicles like trucks and buses use high-speed diesel engines for long-haul efficiency, balancing fuel economy and power for highway driving.

2 Aviation

Light Aircraft Engines – Small planes, like those used in general aviation (e.g., Cessna or Piper), use high-speed piston engines. These engines provide the necessary power to propel smaller aircraft efficiently while keeping weight down.

Jet Engines – Turbofan and turbojet engines are classic examples of high-speed engines used in commercial and military aircraft. These engines have high RPM turbines, providing the necessary thrust for high-speed flight and long-distance travel.

3 Marine

Speedboats and Racing Boats – High-speed marine engines, often powered by inboard or outboard motors, are designed for boats needing quick acceleration and high speeds, such as speedboats, jet skis, and racing boats.

Commercial Vessels – High-speed engines are also used in some ferries, patrol boats, and fishing vessels that require faster transit times, ensuring efficient transportation.

4 Power Generation

Portable Generators – High-speed engines are commonly used in small, portable generators for emergency power supply or in construction sites. These engines provide quick start-up and efficient energy output.

Backup Power Systems – High-speed diesel engines are employed in emergency backup generators for hospitals, data centers, and other critical facilities that require reliable and rapid response during power outages.

Gas Turbines – High-speed gas turbines are used in peak-load power plants to generate electricity quickly during times of high demand, thanks to their ability to ramp up power rapidly.

5 Industrial Applications

Air Compressors – High-speed engines drive industrial air compressors that require high RPM to generate compressed air for tools, machinery, and pneumatic systems in manufacturing.

Pumps – High-speed engines are used to operate industrial pumps for water supply, fire-fighting systems, oil extraction, and chemical processing, providing the necessary force to move fluids efficiently.

Construction Equipment – High-speed engines power heavy machinery such as excavators, bulldozers, and cranes that need reliable engines for quick movement and lifting operations on construction sites.

6 Railway and Transportation

High-Speed Trains – Modern high-speed trains use diesel-electric or electric engines that operate at high speeds to ensure rapid transportation. These engines are optimized for efficiency and high-power output, allowing trains to reach speeds of 300 km/h or more.

Locomotives – High-speed diesel engines are used in locomotives for passenger rail service, where quick acceleration and efficient cruising at high speeds are required.

7 Agriculture

Tractors – In agriculture, high-speed engines are used in tractors and other farm machinery for quick movement across large fields, ensuring timely plowing, seeding, and harvesting.

Harvesters – Combine harvesters and other high-speed farming equipment use engines that provide high torque and RPM to handle tough agricultural tasks efficiently.

8 Military and Defense

Armoured Vehicles – Tanks, armoured personnel carriers, and other military vehicles are equipped with high-speed engines to provide rapid mobility and high power for maneuvering in diverse terrain conditions.

Drones and UAVs – Unmanned aerial vehicles (UAVs) and military drones often use high-speed engines, whether piston or turbine-based, to maintain agility, speed, and prolonged flight capabilities.

9 Motorsports

Formula One – High-speed engines are the heart of Formula One cars, with engines capable of reaching up to 15,000 RPM to achieve incredible acceleration and top speeds. These engines are highly optimized for power, efficiency, and weight reduction.

MotoGP – Similar to Formula One, MotoGP bikes use high-speed engines that offer quick acceleration, agility, and the ability to reach top speeds on race tracks.

10 Oil and Gas Industry

Offshore Drilling – High-speed engines power machinery like pumps, compressors, and cranes used in offshore drilling platforms, where high efficiency and reliability are essential.

Pipeline Operations – High-speed engines are used to drive compressors and pumps in pipeline systems for transporting gas, oil, and other fluids over long distances.

9.1.4 Summary

In summary, High-speed engines play a crucial role in many sectors, providing the necessary power, efficiency, and performance for diverse applications. From powering sports cars and aircraft to driving industrial machinery and generating electricity, these engines are a fundamental component in modern technology and industry. Proper maintenance, condition monitoring, and appropriate application are key to ensuring their reliability and longevity across these demanding environments.

9.2 MAINTENANCE OF HIGH-SPEED ENGINES

9.2.1 Maintenance Schedules for High-Speed Engines

High-speed engines are critical in various applications, from portable power generation to industrial and automotive systems. Regular maintenance ensures these engines operate efficiently, safely, and with a prolonged lifespan. Below is a general outline of maintenance schedules tailored for high-speed engines, addressing daily, periodic, and annual checks.

1 Daily Maintenance

Inspection – Conduct a visual check for oil leaks, fuel leaks, and signs of wear or damage on belts, hoses, and other components.

Oil Levels – Check the engine oil and top up if necessary.

Coolant Levels – Verify coolant levels to avoid overheating.

Air Filter – Inspect the air filter for clogs or dirt; clean or replace as needed.

Operating Parameters – Monitor gauges for oil pressure, temperature, and RPM during operation.

2 Weekly Maintenance

Clean the Engine – Remove accumulated dust and debris to prevent overheating and ensure smooth operation.

Battery Check – Inspect connections and terminals for corrosion; verify charge levels.

Test Run – Operate the engine under light loads to ensure functional components.

Lubrication – Apply lubrication to moving parts, such as fan belts and bearings, as required.

3 Monthly Maintenance

Oil Change – Replace the engine oil and oil filter after a specified number of operating hours (typically 50–100 hours).

Fuel System Inspection – Check fuel filters and replace if clogged; inspect injectors for performance.

Belt Tension – Ensure belts are at proper tension and show no signs of cracks or wear.

Exhaust System – Check for leaks, rust, or blockages in the exhaust system.

4 Quarterly Maintenance

Comprehensive Inspection – Assess all engine components, including spark plugs (if applicable), turbochargers, and cooling systems.

Cooling System Service – Flush and refill the coolant to maintain optimal thermal performance.

Diagnostics – Use diagnostic tools to monitor system performance and detect potential issues.

Fuel Quality – Verify the quality of stored fuel to avoid engine damage.

5 Annual Maintenance

Overhaul: Depending on operating hours, consider a partial or full engine overhaul to replace worn-out components.

Calibration – Calibrate sensors, gauges, and control systems for accurate performance.

Air Intake System – Clean and inspect the air intake system, including turbochargers and manifolds.

Vibration Analysis – Conduct a vibration analysis to identify alignment issues or imbalances.

Load Testing: Run the engine under full load to assess overall efficiency and reliability.

6 Special Considerations

Extreme Conditions – In harsh environments, such as high temperatures or dusty areas, adjust maintenance schedules to more frequent intervals.

Manual Guidelines – Always adhere to the manufacturer's maintenance recommendations for specific engine types.

Record Keeping – Maintain detailed logs of all inspections, repairs, and part replacements for future reference.

9.2.2 Benefits of Scheduled Maintenance

1. **Increased Reliability** – Reduces the likelihood of unexpected breakdowns.
2. **Efficiency** – Maintains optimal performance and fuel efficiency.
3. **Longevity** – Extends the engine's lifespan by minimizing wear and tear.
4. **Cost Savings** – Prevents costly repairs and downtime.

9.2.3 Summary

In summary, regular maintenance is the cornerstone of keeping high-speed engines functioning at their best. Adopting a structured schedule ensures that these engines perform safely and reliably across diverse applications.

CHAPTER TEN

10.0 OPTIMIZING FACTORY LAYOUTS: DESIGN AND CONFIGURATION OF TYPICAL INDUSTRIAL FACILITIES

10.1 METHODS OF PRESENTING LAYOUT

There are multiple methods of presenting layouts with diagrams that cater to different planning, operational, and communication needs. Each type uses specific visual techniques to present a layout's spatial arrangement, workflow, or physical constraints. Several effective methods with corresponding diagrams to illustrate these approaches are illustrated below.

10.1.1 2D Blueprint or Floor Plan

- **Description** – A 2D floor plan (Figure 10.1) provides a top-down view of a space, displaying boundaries, walls, entrances, and major areas within the facility. It often includes equipment and room labels for easy identification.
- **Diagram** – The diagram consists of a simple outline showing rooms, walls, and doors. Each section is labeled, such as "Assembly Area" or "Storage."
- **Usage** – Typically used in architectural and construction planning phases to confirm spatial organization.

Figure 10.1: Sample of 2D Blueprint or Floor Plan

10.1.2 3D Model

- **Description** – A 3D model (Figure 10.2) adds depth to the layout by including heights and relative positioning, allowing for a comprehensive spatial perspective. It's useful for visualizing equipment, pathways, and vertical space.
- **Diagram** – A simplified 3D perspective showing multiple levels, such as shelves, machines, or elevated areas.
- **Usage** – This method is used for complex environments where visualizing height and depth is important, such as warehouses.

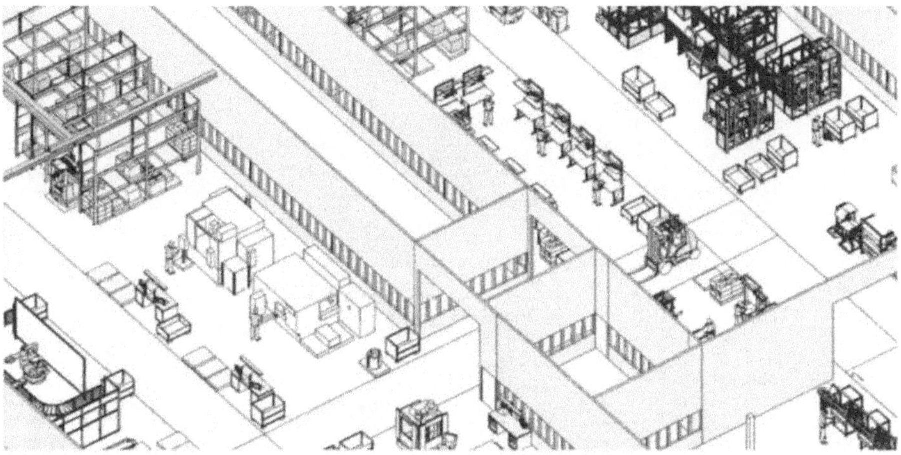

Figure 10.2: Sample of 3D Model Factory Plan

10.1.3 Flow Diagram

- **Description** – Flow diagram (Figure 10.3) focuses on process or movement within the layout. They use arrows to show pathways for products, materials, or people as they progress through stages.
- **Diagram** – Arrows indicate movement, showing directionality between different layout sections.
- **Usage** – Used to optimize workflow and understand the sequence of processes, often in production environments.

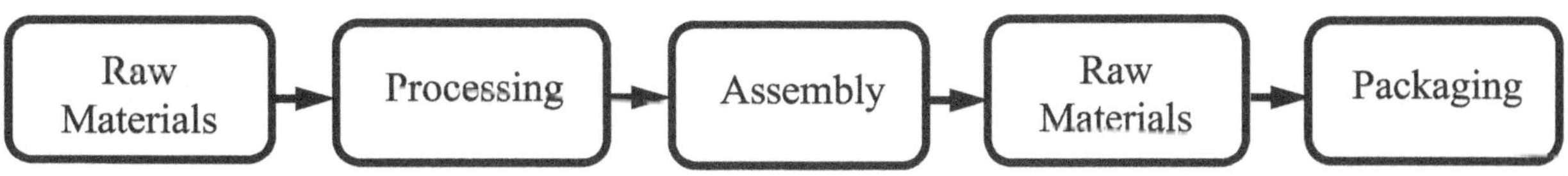

Figure 10.3: Example of Flow Diagram

10.1.4 Spaghetti Diagram

- **Description** – Spaghetti diagram (Figure 10.4) displays actual movement paths within a space, often for efficiency analysis. Lines are drawn from one point to another to represent how individuals or items move around the facility.
- **Diagram** – A collection of crisscrossing lines showing repeated pathways between workstations.
- **Usage** – Useful for identifying bottlenecks and areas with excessive movement to help in layout reconfiguration.

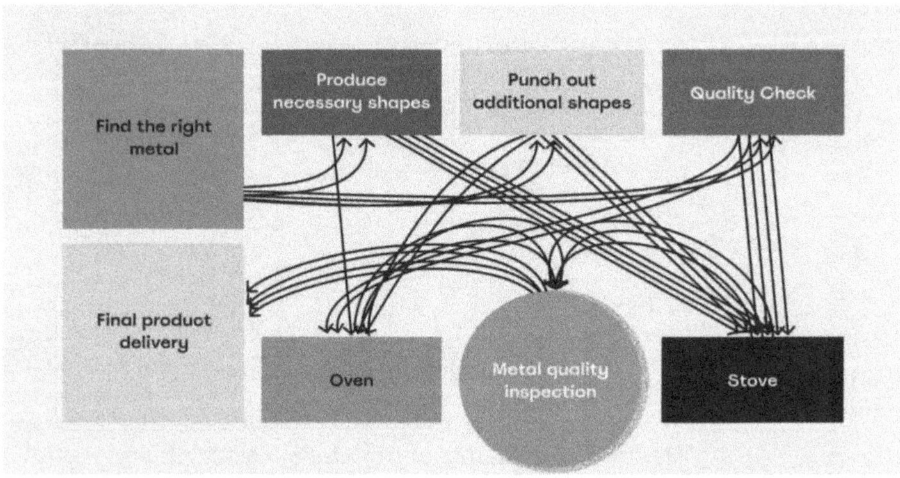

Figure 10.4: Example of Spaghetti Diagram - Mapping out a car-manufacturing factory using a Spaghetti Diagram

10.1.5 Workflow or Process Diagram

- **Description** – Workflow diagram (Figure 10.5) illustrates sequences of actions, typically with boxes or circles for steps and arrows indicating direction. They help show complex processes in a linear or branching format.
- **Diagram** – Each action or stage is represented with a box or shape, and arrows show the flow direction between steps.
- **Usage** – Primarily used to document and optimize processes by clearly displaying each step and its dependencies.

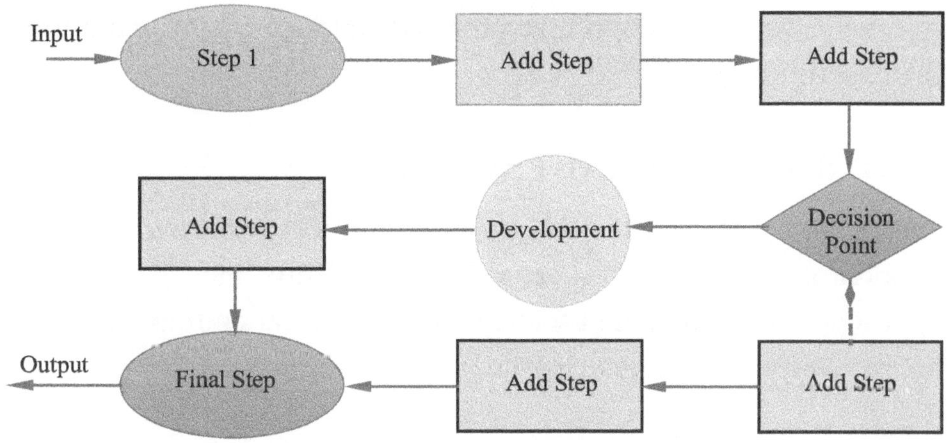

Figure 10.5: Workflow or Process Diagram

10.1.6 Block Diagram

- **Description** – A block diagram (Figure 10.6) represents large sections of a layout or workflow with minimal detail. Each block typically represents a major area or step within the process.
- **Diagram** – Rectangles or squares denote sections or steps, with minimal labeling.
- **Usage** – Useful for high-level views of a facility or process without the complexity of finer details.

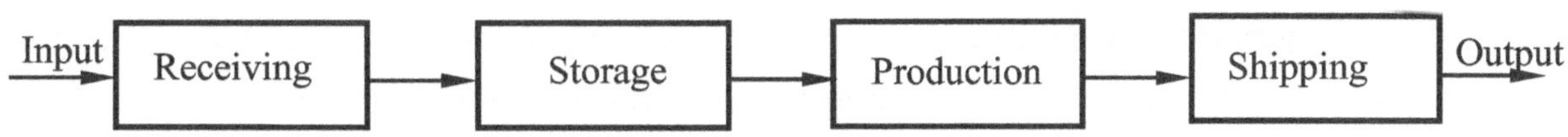

Figure 10.6: Example of Block Diagram

10.1.7 Digital Walkthrough or Virtual Reality (VR) Layout

- **Description** – Digital walkthroughs or virtual reality layouts offer a realistic, interactive view of a space that can be navigated digitally. This method allows stakeholders to experience the layout first-hand.
- **Diagram** – Screenshots or stills from a virtual view, such as images from a 360-degree camera or VR environment.
- **Usage** – Ideal for training and reviewing facility layouts before physical construction or installation, especially for safety assessments.

Each diagram method provides a unique level of insight into the layout, suited to specific planning needs, from high-level overviews to detailed spatial planning. Using these visual approaches helps companies optimize processes, reduce errors, and improve overall efficiency within their facilities.

10.2 LAYOUT OF TYPICAL FACTORIES

The layout of typical factories is designed to optimize workflow, minimize material handling, ensure worker safety, and improve efficiency. While the specific layout depends on the industry, production method, and scale, there are several common types that serve as effective blueprints for various manufacturing needs.

10.2.1 Process Layout (Functional Layout)

Description – Created by arranging employees in groups based on similar tasks and skills. The process layout is also known as functional layout, where ~~In a process layout,~~ machines and equipment are grouped by the function they perform (e.g., all milling machines are together, all assembly stations are together).

Ideal For – Factories producing customized or low-volume products, such as workshops or machine shops.

Advantages: High flexibility, equipment can be used for different products, and employees develop specialized skills in specific tasks.

Example – A factory where welding, painting, and assembling departments are in different areas, with each department handling parts before passing them to the next stage. Figure 10.7 shows a typical manufacturing process layout which the various workstations are indicated by letters.

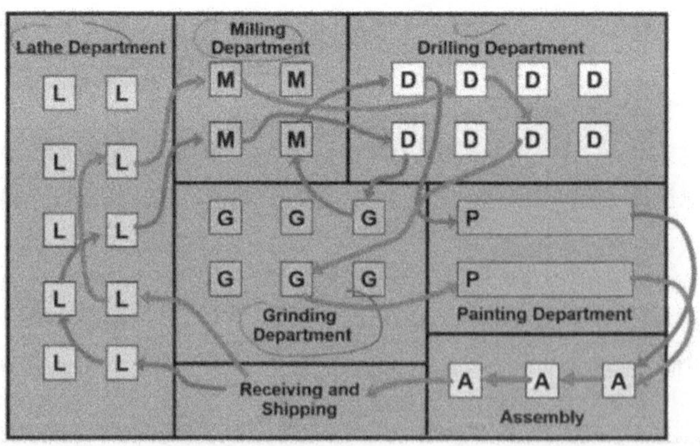

Figure 10.7: Manufacturing Process Layout

Figure 10.8 illustrates plant layout consisting of four operational sections: Cutting Section – Bending Section – Machining Section – Welding Section which the workstations are indicated by numbers.

Each section has a certain amount of specialized machinery and equipment, similar in their operational application:

- Cutting Section - cutting machines: 1 - 2 - 3
- Bending Section - bending brakes: 4 - 5 - 6
- Machining Section - NC machines: 7 - 8 - 9 - 10
- Welding Section - arc welders: 11 - 12 - 13

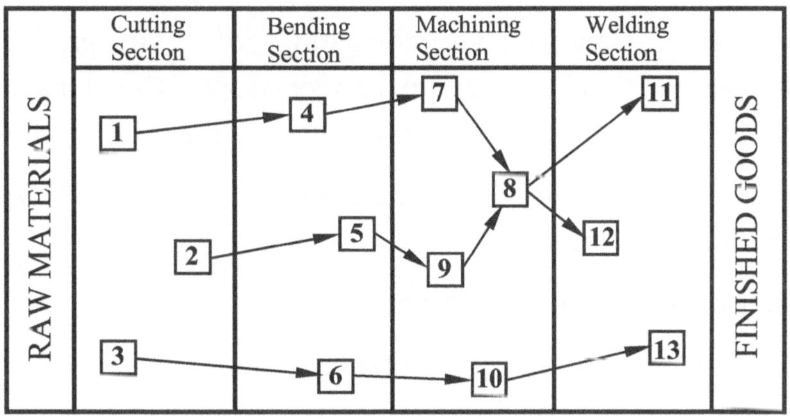

Figure 10.8: Process Layout

10.2.2 Product Layout (Assembly Line Layout)

Description – This layout arranges equipment in a sequential order based on the steps in the manufacturing process. It is linear and follows a specific sequence, often resembling a conveyor belt or assembly line.

Ideal For – High-volume production with little variation, such as automotive manufacturing or electronics assembly.

Advantages – High efficiency for standardized production, reduced material handling, and ease in monitoring workflow.

Example – An automotive plant where each station on the line performs a specific task (e.g., installing the engine, attaching wheels) as the car moves down the line.

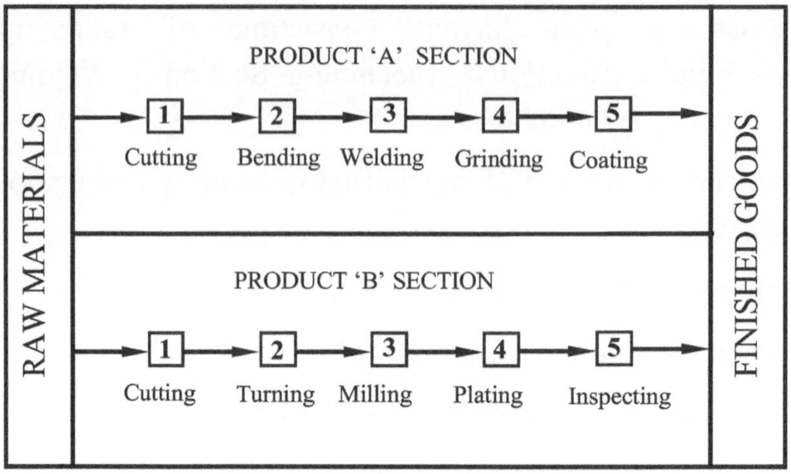

Figure 10.9: Product Layout

Figure 10.9 illustrates plant layout consisting of two product sections, **Product A** and **Product B.** Each section has a certain number of specialized machinery and equipment, which provide all operations required by a particular manufacturing process.

The manufacturing process of Product A, for example, consists of five specialized operations, represented in the following order:

1 Cutting of sheet metal - cutting machine.

2 Bending of sheet metal parts - bending brake.

3 Welding various components together - spot welder.

4 Grinding of the assembled product - grinder.

5 Coating of the assembled product - dry powder equipment.

10.2.2.1 Advantages of Product Layout

1 It provides a steady flow of work over direct operational routes, minimizing delays in the manufacturing process.

2 Less material handling because of shorter travel of work between relevant working stations.

3 It provides for closer coordination of the manufacturing process as a result of the pre-planned operational sequence.

4 It helps to minimize manufacturing time per unit of product.

5 It helps to reduce the volume of work-in-process between various working stations.

6 It requires smaller floor areas per unit of product within the production facility.

7 It simplifies production planning and control procedures since all operations are synchronized well in advance.

10.2.2.2 Disadvantages of Product Layout

1 There is a need for higher capital investment in machinery and equipment because of duplication of some manufacturing processes.

2 There is an increased potential for machine idleness if certain production lines are not fully loaded to their operational capacity.

3 There is less flexibility in the process of work accomplishment because jobs cannot be re-routed to other similar machines as in the process layout arrangement.

4 There is little development of workers' operating skills because each operator only learns his or her job at a specific machine or working station. This results in decreased universal abilities of workers in operating a range of machinery and equipment.

5 There is no highly efficient supervision because supervisors and foremen become too involved in specific manufacturing processes and neglect the overall in-depth knowledge required to set up and regulate the operation of various machinery and equipment. This process requires skilled setters to prepare specific machines for work in order to meet manufacturing process requirements.

6 There is an overall increase in manufacturing costs, although labor costs per unit of product may be lower. This increase is a direct result of high facility overheads on production lines, particularly if the utilization of the operational capacity is not used fully, and if certain machines are idle.

7 There is an increased danger of production line shutdown if certain machinery or equipment fails to perform. It is necessary, therefore, to provide standby equipment or additional manufacturing facilities in order to prevent unnecessary production delays and to keep the work flowing.

10.2.3 Cellular Layout

Cellular Layout is a layout strategy used in manufacturing and production environments where machines, equipment, and workstations are grouped into "cells." Each cell is designed to perform a specific sequence of operations necessary to manufacture a part or product family. The key characteristic of a cellular layout is that it groups together similar processes or parts with similar processing requirements, allowing for a more efficient and streamlined production process.

This type of layout aims to create a small, self-contained production unit that operates like a mini production line within a larger facility. Cellular layouts are commonly used in industries where flexibility and efficiency are crucial, particularly in batch or small-to-medium volume production.

10.2.3.1 Key Features of Cellular Layout

1 **Product Family Focus** – Machines and equipment are grouped into cells based on the product family or parts with similar processing needs.

2 **Reduced Movement** – Minimizes the movement of materials and products by keeping the production steps within a cell.

3 **Flexibility** – Allows for quick reconfiguration of cells to adapt to changes in product demand or design.

4 **Continuous Flow** – Promotes a continuous flow of work, reducing waiting times and inventory between operations.

10.2.3.2 Advantages of Cellular Layout

1 **Efficiency** – Reduces material handling time and increases production speed due to the close proximity of related processes.

2 **Reduced Setup Time** – Equipment within a cell is often arranged to minimize changeovers, leading to faster setup times.

3 **Better Quality Control** – Easier to monitor and maintain quality within a small production cell.

4 **Flexibility** – Quick to adapt to changes in product design or customer demand.

5 **Employee Satisfaction** – Often involves cross-training workers to handle various tasks within a cell, increasing job satisfaction.

10.2.3.3 Disadvantages of Cellular Layout

1 **Initial Setup Cost** – Requires investment in grouping machines into cells and may need specialized equipment.

2 **Training** – Workers need to be multi-skilled to handle different tasks within a cell, requiring additional training.

2. **Space Constraints** – May require more space than traditional layouts if cells are not properly designed.

3. **Underutilization** – If production volume is low, some equipment within a cell may remain idle.

10.2.3.4 Examples of Cellular Layout

1 Automobile Manufacturing

Example – In automotive manufacturing, a cellular layout might be used for the assembly of specific components like engines or transmission systems. For instance, one cell could be responsible for assembling engine parts—cylinder blocks, pistons, and crankshafts—while another cell focuses on the transmission assembly. Each cell operates independently, completing a full assembly of a component before moving it to the main assembly line.

2 Electronics Manufacturing

Example – In electronics production, companies like Panasonic or Sony use cellular layouts for assembling circuit boards. A cell might include machines and workers responsible for mounting components, soldering, and testing a family of similar circuit boards. This reduces the need to transport partially completed boards across the facility, enhancing efficiency.

3 Furniture Production

Example – A furniture manufacturing company producing a line of office desks and chairs can set up a cell for cutting, assembling, and finishing wooden parts. Each cell is designed to handle the entire manufacturing process for one type of furniture, reducing the time it takes to move parts from one stage to another.

4 Medical Device Manufacturing

Example – In the production of medical devices like surgical instruments, a cell might be dedicated to producing a specific product family. For example, a cell could handle the entire process of manufacturing orthopedic screws, from raw material shaping to final polishing and quality inspection.

5 Aerospace Manufacturing

Example – Aerospace manufacturers use cellular layouts to build specific components of an aircraft, like wings or fuselage sections. Each cell is responsible for assembling a specific component, and workers within the cell are cross-trained to perform multiple tasks, allowing for flexibility in production.

6 Metal Fabrication

Example – In a metalworking plant, a cell might be created to manufacture a series of metal brackets. The cell includes cutting, bending, welding, and finishing equipment arranged in a sequence. By using a cellular layout, the company reduces the movement of parts between different departments, speeding up production.

10.2.3.5 When to Use Cellular Layout

Cellular Layout (Figure 10.10) is ideal for environments where:

- **Product Variety is High**, and products can be grouped into families with similar characteristics.
- **Production Volume is Moderate** to low, allowing for flexibility without the rigidity of a traditional assembly line.
- **Lean Manufacturing Goals** are in place, aiming to minimize waste and improve flow.

The cellular layout is a core strategy in lean manufacturing practices, as it helps reduce waste, improve efficiency, and promote flexibility. It supports industries that require high levels of customization, batch production, and frequent design changes while maintaining efficiency in the manufacturing process.

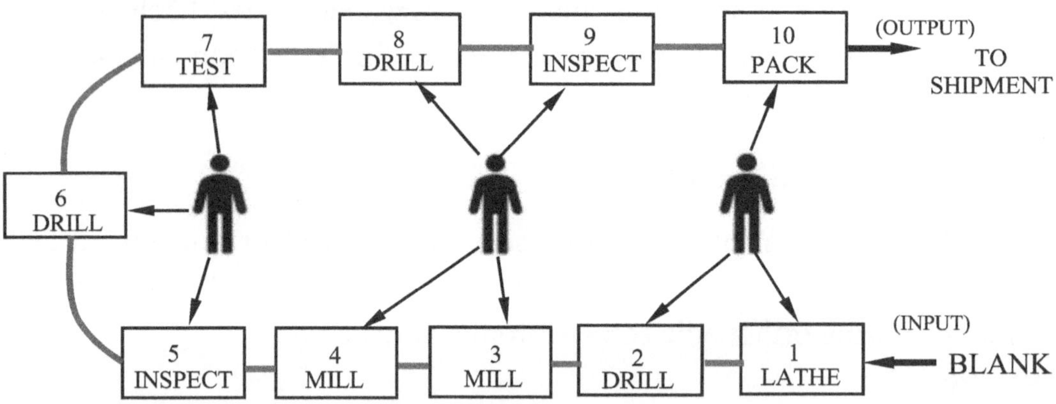

Figure 10.10: Cellular Layout

10.2.4 Fixed-Position Layout

A **Fixed-Position Layout** (Figure 10.11 and Figure 10.12) is a facility layout strategy where the product being manufactured or assembled remains in a fixed location, and all necessary resources—workers, materials, tools, and equipment—are brought to the site. This layout is particularly useful for products that are bulky, heavy, or too large to move easily, making it more efficient to bring the necessary components to the worksite rather than moving the product through a production line.

10.2.4.1 Key Features of Fixed-Position Layout

1 **Stationary Product** – The main feature of a fixed-position layout is that the product does not move during the production process. All activities occur around a fixed point.

2 **Resource Movement** – Workers, machinery, and materials are moved to the location where the product is being worked on.

3 **Complexity and Scale** – It is suitable for large-scale or complex projects that involve multiple phases and specialized teams.

4 **Customization** – Allows for greater customization and flexibility, as workers can focus on specific tasks without the constraints of a production line.

10.2.4.2 Advantages of Fixed-Position Layout

1 **Reduced Handling** – Minimizes the movement of heavy or large products.

2 **Customization**: Easier to modify or adapt to customer-specific requirements.

3 **Flexibility**: Different teams can work simultaneously on different parts of the product.

10.2.4.3 Disadvantages of Fixed-Position Layout

1 **Coordination Challenges** – Requires effective coordination between teams to avoid delays.

2 **Limited Space** – Can become crowded as various workers and equipment are concentrated in one area.

3 **Higher Cost** – Typically, more labor-intensive, leading to higher costs compared to other layouts.

10.2.4.4 Examples of Fixed-Position Layout

1 **Shipbuilding**

Example – In a shipyard, the construction of a large ship like a cruise liner or cargo vessel takes place in a dry dock or a specific area. The ship remains stationary while different teams handle hull assembly, engine installation, electrical systems, and interior finishing. Moving a partially constructed ship would be impractical, so the fixed-position layout allows for simultaneous operations on different parts of the ship.

2 Aircraft Assembly

Example – Aircraft manufacturers like Boeing and Airbus utilize a fixed-position layout for assembling airplanes. An aircraft remains in a large hangar while workers install engines, wings, avionics, interiors, and other components. Each stage requires specialized teams, and moving an aircraft during construction would be inefficient and costly.

3 Building Construction

Example – In construction projects, such as building skyscrapers, houses, or bridges, the construction site is the fixed point. All materials, cranes, equipment, and workers come to the site. For instance, during the construction of a high-rise building, different teams work on the foundation, structure, electrical wiring, plumbing, and interior finishing, all within the same location.

4 Power Plant Installation

Example – Constructing a power plant requires heavy machinery, large equipment, and specialized labor. Equipment like turbines, generators, and boilers are installed on-site. The size and complexity of the equipment necessitate a fixed-position layout, as it would be nearly impossible to move them around.

5 Rocket Assembly

Example – Space agencies like NASA or SpaceX use fixed-position layouts for assembling rockets. The rockets are constructed in large hangars or on launch pads. Various teams work on different sections—fuel tanks, engines, payload sections—while the rocket remains in one location. It simplifies the assembly of massive components and reduces the risk associated with moving them.

6 Custom Furniture Manufacturing

Example – For custom-built furniture, such as a large wooden conference table, the product might remain in a workshop or assembly area while artisans bring different components like wood, veneers, tools, and finishing materials to the workstation. This setup is ideal when precision and artisanship are required.

10.2.4.4.1 When to Use Fixed-Position Layout

Fixed-Position Layout is best suited for projects involving:

1. **Large and Heavy Products** that are difficult to move.
2. **Complex and Multi-Stage Processes** requiring specialized teams.

3 **Customization** where each product needs to meet unique specifications.

This layout emphasizes efficient resource movement and effective coordination of labor and equipment, making it suitable for industries where the scale and complexity of projects require a stationary product.

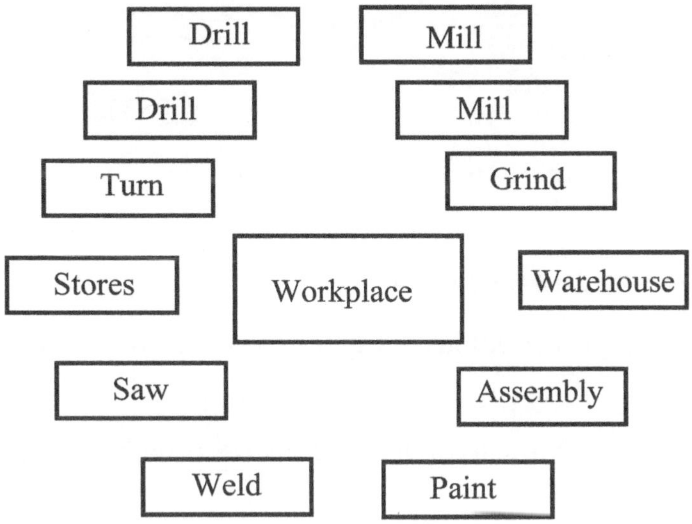

Figure 10.11: Block Diagram of Fixed-Position

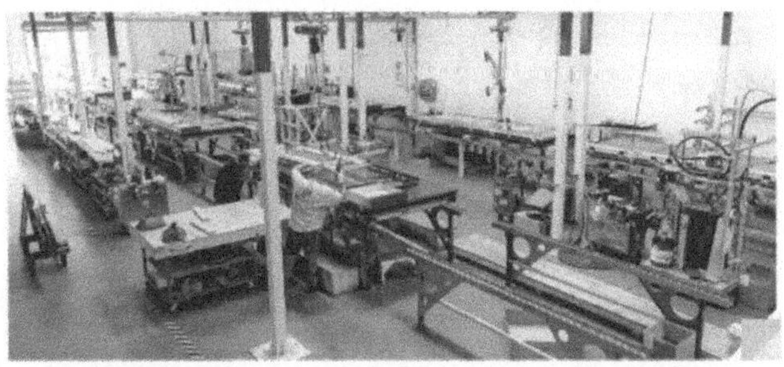

Figure 10.12: Picture of Fixed-Position

10.2.5 Hybrid Layout (Combination Layout)

Hybrid Layout (an example shown in Figure 10.13), also known as a Combination Layout, is a facility layout strategy that blends elements from multiple layout types, such as process, product, and fixed-position layouts, to create a system that optimally meets the unique needs of a particular production environment. The goal of a hybrid layout is to leverage the advantages of various layout types while minimizing their individual limitations.

Hybrid layouts are widely used in modern industries to handle complex manufacturing environments, where a single layout type may not be suitable for all production needs. This approach enables companies to achieve higher levels of flexibility, efficiency, and scalability.

Figure 10.13: Example of Hybrid Layout

10.2.5.1 Key Features of Hybrid Layout

1 **Flexibility** – Combines different layout types to accommodate a diverse range of products or processes.

2 **Optimized Workflow** – Aims to enhance the workflow by placing equipment, machines, and workstations strategically.

3 **Customizability** – The layout can be easily adapted to changes in production demand, design, or new product introduction.

4 **Cost Efficiency** – Reduces unnecessary movements and handling, leading to cost savings in material handling and production.

10.2.5.2 Advantages of Hybrid Layout

1 **Versatility** – Suitable for a wide range of products and processes.

2 **Increased Efficiency** – Combines the strengths of multiple layouts to improve overall productivity.

3 **Improved Space Utilization** – Can be optimized for better use of available space.

4 **Ease of Supervision** – Enhances the ability to manage different processes effectively.

5 **Greater Adaptability** – Easily adaptable to production changes or expansion.

10.2.5.3 Disadvantages of Hybrid Layout

1 **Complex Planning** – Designing a hybrid layout requires careful planning and expertise.

2 **Higher Initial Setup Cost** – May involve higher upfront investment due to specialized equipment and infrastructure.

3 **Potential Congestion** – If not well designed, different layout elements can create congestion and bottlenecks.

4 **Maintenance Challenges** – A diverse setup may increase maintenance complexity.

10.2.5.4 Examples of Hybrid Layout

1 Automotive Assembly Plant

Example – In automotive manufacturing, companies like Toyota use hybrid layouts to combine product layout (assembly line for car assembly) and process layout (workshops for specialized tasks like painting or engine testing). The main assembly line follows a product layout for efficient mass production, while certain specialized areas operate in a process layout to handle diverse customization options.

2 Electronics Manufacturing

Example – An electronics manufacturer might use a hybrid layout that blends process layout (where similar operations like soldering and testing are grouped) and cellular layout (small cells for specific product families such as motherboards or power units). This approach allows efficient handling of different types of electronic devices while accommodating variations in volume and design.

3 Furniture Production Facility

Example – A company producing custom furniture may use a combination layout that incorporates fixed-position layout for assembling large, immovable pieces (e.g., custom tables) and process layout for crafting smaller components like chair legs. By merging the two, the company can handle both mass production and custom orders efficiently.

4 Hospital Layout

Example – Hospitals often use a hybrid layout by combining departmental layouts (like separate areas for radiology, surgery, and emergency) with fixed-position layouts (operating rooms where surgical equipment remains stationary). This enables the hospital to manage a wide range of services while maintaining efficiency and specialization.

5 Aircraft Manufacturing

Example – In the aerospace industry, companies like Boeing employ hybrid layouts by integrating fixed-position layout (where aircraft remain stationary during assembly) and cellular layouts for specific parts like wings or engines. This allows for the efficient assembly of large, complex aircraft with specialized sub-assembly cells for individual components.

6 **Retail Store Layout**

Example – Retailers like IKEA utilize a hybrid layout by combining product layout (aisle organization) with fixed-position layout for demonstration areas or mock room setups. This allows customers to navigate efficiently through product displays while also experiencing how items might look in a home setting.

10.2.5.6 When to Use Hybrid Layout

Hybrid Layout is ideal in situations where:

1. A diverse range of products needs to be produced, requiring different processing approaches.
2. Production involves both standardized products and custom-made items.
3. There is a need to accommodate high volume and low volume production in the same facility.
4. A company wants to achieve a balance between flexibility and efficiency, adapting to market demands.

10.2.5.7 How to Design a Hybrid Layout

1. **Identify Core Production Needs** – Determine the primary types of products or processes that need to be supported.
2. **Analyse Workflow** – Study how materials, products, and information move through the facility to identify potential bottlenecks or inefficiencies.
3. **Combine Layout Types** – Choose elements from product, process, fixed-position, or other layouts based on production requirements.
4. **Test and Adjust** – Use simulations or pilot runs to identify any challenges or areas for improvement.
5. **Implement and Monitor** – Deploy the hybrid layout and monitor its performance, making adjustments as necessary to optimize efficiency.

10.2.5.8 Summary

In summary, a hybrid layout is a powerful strategy for manufacturing environments that face diverse production demands and frequent changes. By blending different layout types, companies can create a flexible, efficient, and scalable production system tailored to their unique needs. The success of a hybrid layout depends heavily on careful planning, clear workflow analysis, and ongoing optimization.

10.2.6 Warehouse or Storage Area

A Warehouse or Storage Area is a designated space used to store raw materials, work-in-progress items, and finished goods before they are distributed or used. These areas are a crucial part of the supply chain, helping companies maintain inventory, organize materials, and manage logistics. Properly designed and maintained storage areas ensure the safe and efficient handling of goods, reduce operational costs, and play a key role in optimizing productivity.

10.2.6.1 Key Functions of a Warehouse or Storage Area

1 **Inventory Storage** – Warehouses provide a space to store products or raw materials securely until they are needed in production or ready for distribution.

2 **Inventory Management** – Storage areas enable systematic inventory tracking, ensuring accurate stock levels, efficient restocking, and easy retrieval of goods.

3 **Quality Control** – Warehouses can serve as quality inspection points where goods are checked and sorted before being moved to the next phase in the supply chain.

4 **Distribution Hub** – Warehouses often act as distribution centers, facilitating the transportation of goods to retailers, end customers, or other locations.

5 **Order Fulfillment** – Warehouses house order fulfilment processes, including picking, packing, and shipping goods.

10.2.6.2 Types of Warehouses or Storage Areas

1 **Raw Material Warehouse** – Stores raw materials and components required for manufacturing.

Example – A **car manufacturing plant** might have a raw material warehouse for steel sheets, car components, electronics, and other materials that are essential for the assembly line.

2 **Work-In-Progress (WIP) Warehouse** – Stores partially completed goods that are waiting to be moved to the next production phase.

Example – In a **furniture factory**, a WIP storage area may hold unfinished wooden parts, such as sanded chair legs, that need further processing before final assembly.

3 **Finished Goods Warehouse** – Holds completed products that are ready for sale or distribution.

Example – A **textile company** might use a finished goods warehouse to store clothes, curtains, and textiles that are ready to be shipped to retail stores.

4 **Distribution Center** – Specialized warehouses designed to receive, process, and ship products to different destinations.

Example – **Amazon's fulfilment centers** are a classic example of distribution centers, handling millions of products from multiple suppliers, packaging them, and shipping them directly to customers.

5 **Cold Storage Warehouse** – Used to store perishable goods, such as food and medical supplies that require temperature control.

Example – A **grocery chain** may have cold storage facilities for frozen goods like seafood, dairy products, and vegetables to maintain product quality and extend shelf life.

6 **Bulk Storage Warehouse** – Designed for storing items in bulk quantities, often in large containers or on pallets.

Example – A **construction material supplier** might use bulk storage for concrete bags, sand, gravel, and large amounts of construction materials.

7 **Automated Warehouse** – Equipped with automation technologies such as robotic systems, conveyor belts, and warehouse management software (WMS) for efficient handling of inventory.

Example – **Zappos**, a major online shoe and clothing retailer, employs automated warehouses to handle thousands of product lines using conveyor systems and automated storage and retrieval systems (ASRS).

10.2.6.3 Examples of Warehouse Layouts

1 **Straight-Line Layout** – Products move in a straight line from receiving to storage, picking, packing, and shipping as shown in Figure 10.14.

Example – Commonly seen in **e-commerce warehouses**, where efficiency in picking and packing is crucial.

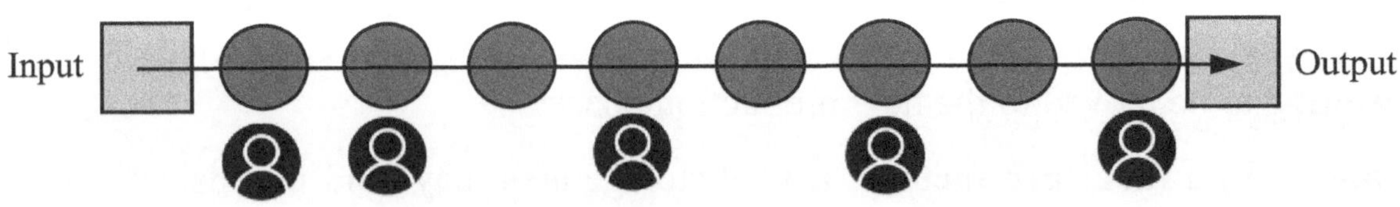

Figure 10.14: Example of Straight-Line Layout

2 **U-Shaped Layout** (Figure 10.15)– Receiving and shipping areas are located next to each other, with storage in the middle. It optimizes space and minimizes travel time for workers.

Example – Used by **grocery store warehouses**, facilitating easy access to both incoming goods and outgoing orders.

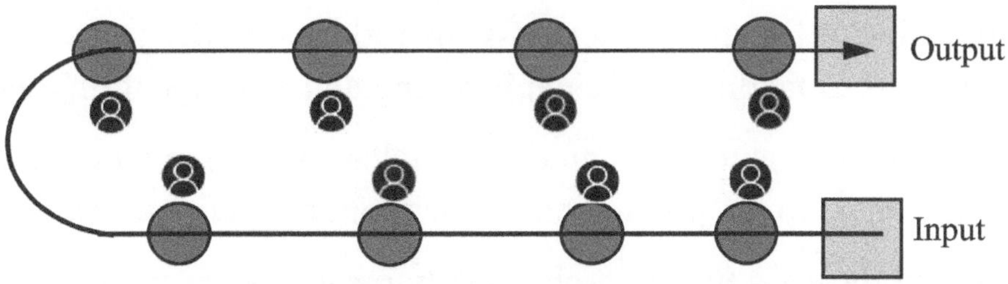

Figure 10.15: Example of U-Shaped Layout

3 **L-Shaped Layout** (Figure 10.16) – Similar to the U-shaped layout, but with a right-angle configuration. Suitable for warehouses with limited space.

Example – Used in **pharmaceutical warehouses** to separate receiving and dispatching areas clearly.

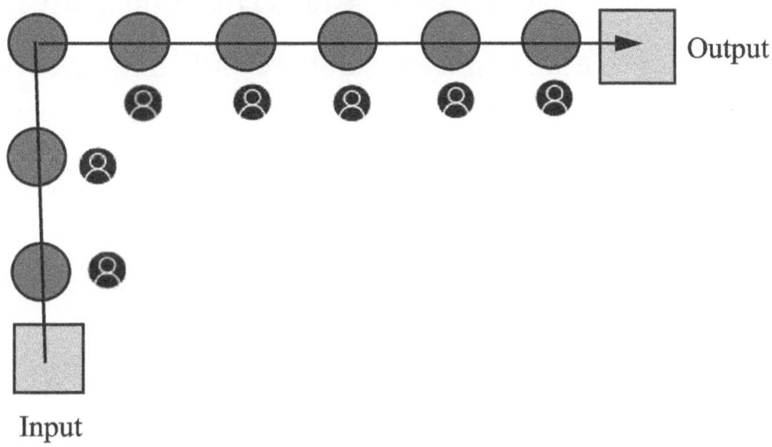

Figure 10.16: Example of L-Shaped Layout

10.2.6.4 Key Considerations for Effective Warehouse Management

1 **Space Utilization –** Use racks, shelves, and bins to maximize storage capacity. Properly organized spaces improve inventory visibility and reduce retrieval times.

Example – In a **clothing warehouse**, vertical storage racks for hanging garments can increase capacity while keeping items wrinkle-free.

2 **Inventory Control**: Implement effective inventory management systems to track stock levels, reduce overstock, and prevent shortages. Barcode systems, RFID tags, and Warehouse Management Systems (WMS) are commonly used.

Example – A **medical supply warehouse** might use RFID technology to track critical supplies, ensuring quick access in emergencies.

3 **Material Handling** – Equip warehouses with forklifts, pallet jacks, conveyor systems, and robotic systems to facilitate efficient handling and movement of goods.

Example – Toyota uses automated guided vehicles (AGVs) in their warehouses to move materials from storage to the production line without manual intervention.

4 **Safety and Compliance** – Ensure proper safety measures are in place to protect workers and goods. This includes safe storage practices, proper lighting, safety equipment, and adherence to safety regulations.

Example – A **chemical warehouse** must follow strict guidelines for storing hazardous materials, such as using spill containment pallets and maintaining proper ventilation.

10.2.6.5 Challenges in Warehouse Management

1 **Space Constraints** – Limited space can lead to congestion, affecting efficiency.

2 **Inventory Accuracy** – Maintaining accurate records of stock levels can be challenging without proper systems in place.

3 **Handling Returns** – Managing returned items requires separate areas and processes, complicating the layout.

5 **Seasonal Fluctuations** – Seasonal demand can lead to sudden spikes or drops in inventory, necessitating flexible storage strategies.

10.2.6.6 Summary

In summary, warehouses or storage areas play a pivotal role in supply chain efficiency, impacting everything from inventory management to distribution logistics. Effective warehouse management requires careful planning of layout, organization of storage space, safety measures, and proper use of technology. Different industries have unique needs, and the type of warehouse or storage area should be tailored to the specific requirements of the business. Properly designed and managed storage areas not only enhance productivity but also reduce costs, improve inventory accuracy, and ensure the safe handling of goods. Figure 10.17 shows an example of warehouse.

Figure 10.17: Example of Warehouse

10.2.6.7 Administration and Office Area

The **Administration and Office Area** is a dedicated space in industrial and commercial settings where administrative functions, managerial activities, and various office-related tasks are conducted. This area is crucial for ensuring that the operational side of a business is well-coordinated, documented, and managed. It serves as the central hub for communication, strategic planning, financial management, human resources, and other essential business functions. A well-organized administration and office area supports the efficiency and productivity of the entire organization.

10.2.6.8 Key Functions of Administration and Office Areas

1 **Management and Decision-Making** – These areas house key personnel involved in strategic planning, decision-making, and overseeing the daily operations of the business.

Example – A manufacturing company's head office might be where managers and supervisors plan production schedules, set business goals, and make key decisions.

2 **Human Resources (HR) Management** – Handling employee-related tasks, such as recruitment, training, payroll, performance evaluation, and maintaining employee records.

Example – In a large construction firm, the HR office may manage hiring skilled labor, organize training sessions, and handle employee benefits.

3 **Financial Management** – Financial tasks, including budgeting, accounting, invoicing, payroll processing, and financial planning, are handled in this area.

Example – In a retail business, the finance office is responsible for tracking inventory costs, managing cash flow, and ensuring timely payment of suppliers.

4 **Communication Hub** – Administration offices often serve as communication centers, facilitating interaction between different departments, customers, and stakeholders.

Example – In a logistics company, the office area might coordinate communication between the warehouse, transportation, and external clients to ensure timely deliveries.

5 **Document Storage and Record Keeping** – Offices maintain important documentation, contracts, employee files, and business records in a well-organized manner, ensuring quick retrieval when needed.

Example – A healthcare facility's administration area may store patient records, insurance documents, and compliance files securely, ensuring privacy and data protection.

6 **Customer Service** – Handling inquiries, managing orders, and providing support to clients are key customer service tasks conducted in office areas.

Example – In an e-commerce company, the customer service team handles phone inquiries, manages online chat support, and processes returns and complaints.

10.2.6.9 Types of Administration and Office Areas

1 **Reception Area** – The first point of contact for visitors, where inquiries are handled, and guests are directed to appropriate personnel.

Example – In a law firm, the reception area acts as the waiting room for clients, with receptionists handling appointments and directing visitors.

2 **Executive Offices** – These are private offices for high-level managers and executives responsible for strategic decisions and overall company leadership.

Example – In a tech company, the CEO's office is a dedicated space for high-level decision-making, strategy meetings, and executive communication.

3 **Open Office Layout** – A common office layout with shared workspaces that promote collaboration, communication, and efficient use of space.

Example – In a software development company, developers and project managers may work in an open-plan office to foster teamwork and facilitate quick communication.

4 **Conference Rooms** – Designated spaces for meetings, discussions, presentations, and team collaboration. Equipped with audiovisual tools, projectors, and video conferencing systems.

Example – In a consulting firm, conference rooms are used to host client meetings, team discussions, and project presentations.

5 **Cubicles**: Divided workspaces for individual employees, providing privacy while allowing communication between team members.

Example – In a call center, employees work in cubicles, ensuring they have enough space to handle calls while minimizing noise from colleagues.

6 **Break Room or Pantry** – A dedicated space where employees can relax, eat, or take breaks during working hours.

Example – A large manufacturing plant might have a break room where employees can have lunch, equipped with a microwave, coffee machine, and comfortable seating.

10.2.6.10 Key Considerations for Designing Administration and Office Areas

1 **Space Utilization** – Efficient use of space is critical to accommodate all necessary administrative functions without overcrowding.

Example – An engineering firm's office may use modular desks that can be rearranged as needed, optimizing the use of available space.

2 **Ergonomics** – Ensuring comfortable furniture and proper lighting to promote employee well-being and reduce strain.

Example – In a financial consulting company, ergonomic office chairs and height-adjustable desks may be used to reduce back and neck strain for employees.

3 **Technology Integration** – Equipping the office with the latest technology tools like computers, telephones, networking equipment, and video conferencing systems to support daily operations.

Example – A marketing agency may use a range of devices, including desktop computers, high-speed internet, and projectors, to deliver digital presentations and manage social media campaigns.

4 **Security and Access Control** – Offices may require security systems like keycard access, biometric scanners, and surveillance cameras to protect sensitive information and maintain privacy.

Example – A pharmaceutical company's administration office may have restricted access, allowing only authorized personnel to enter the areas dealing with research data and intellectual property.

5 **Storage and Filing** – Providing adequate storage solutions like filing cabinets, shelving units, and document safes for record keeping and inventory management.

Example – An architectural firm's office may have specialized filing cabinets for blueprints, technical drawings, and project files.

10.2.6.11 Examples of Administration and Office Layouts

1 **Traditional Office Layout** – Features individual rooms or cubicles for employees, providing privacy and dedicated workspaces.

Example – A legal office might use a traditional layout, with each attorney having a private office for confidential client consultations.

2 **Modern Open Office Layout** – Uses open spaces to encourage collaboration, flexibility, and easy communication between employees.

Example – A creative design studio may use an open office with shared tables, comfortable seating, and whiteboards to foster a collaborative environment.

3 **Hybrid Office Layout** – Combines open spaces and private offices to offer flexibility, allowing both collaboration and privacy when needed.

Example – A tech start-up might use a hybrid layout, with open areas for developers and private meeting rooms for confidential discussions.

10.2.6.12 Challenges in Managing Administration and Office Areas

- **Noise Levels** – Open office layouts can lead to noise distractions that hinder productivity.
- **Space Constraints** – Limited space may affect comfort and the ability to accommodate all functions efficiently.
- **Security Concerns** – Handling sensitive data and ensuring privacy can be challenging without proper security measures.
- **Technology Dependency** – Office work often relies heavily on technology, requiring regular updates and maintenance to prevent disruptions.

10.2.6.13 Summary

The Administration and Office Area (Figure 10.18) plays a vital role in supporting the operational, managerial, and strategic functions of an organization. The design and functionality of these spaces significantly influence productivity, employee well-being, and the overall efficiency of the business. Properly planned and equipped office areas ensure streamlined processes, effective communication, and a positive work environment, ultimately contributing to the success of the organization.

Figure 10.18: Open Plan Administration Office

10.2.6.14 Quality Control and Inspection Area

The **Quality Control (QC) and Inspection Area** is a designated space within industrial facilities, manufacturing plants, and production units where products are systematically checked and verified to ensure they meet predefined standards. The goal of quality control is to maintain high standards, eliminate defects, and ensure customer satisfaction by delivering reliable and high-quality products. This area plays a crucial role in identifying errors, ensuring consistency, and maintaining compliance with industry-specific regulations.

10.2.6.15 Purpose of the Quality Control and Inspection Area

1 **Verification of Product Quality** – Ensure that all products meet the required specifications and standards before they are moved to the next stage of production or released to the market.

   Example – In a car manufacturing plant, the QC area checks each vehicle for compliance with safety standards, design specifications, and overall functionality.

2 **Identification of Defects** – Detect defects or deviations from product specifications early to prevent them from reaching the customer or causing larger production problems.

   Example – A pharmaceutical company's QC lab may identify and reject batches of medicine that do not meet purity and potency standards.

3 **Compliance with Regulations** – Ensure that products comply with industry and safety standards set by regulatory bodies. This compliance is often mandatory to avoid legal issues and product recalls.

   Example – In a food processing plant, the QC area monitors adherence to food safety guidelines, such as checking for contaminants and ensuring proper packaging.

4 **Process Improvement** – Analyse data collected from inspection results to identify trends and areas for improvement in production processes.

Example – In an electronics manufacturing facility, data from QC checks on faulty circuits may lead to changes in the soldering process to improve quality.

10.2.6.16 Key Activities in the Quality Control and Inspection Area

1 **Sampling and Testing** – Conducting tests and measurements on a sample of products from a batch to check for defects and ensure compliance with specifications.

Example – In a textile factory, random fabric samples are taken to test for color consistency, tensile strength, and durability.

2 **Visual Inspection** – Physically examining products to identify visible defects such as cracks, deformations, or misalignments.

Example – In a glass manufacturing plant, workers visually inspect each glass sheet for cracks, air bubbles, or scratches.

3 **Dimensional Inspection** – Measuring the dimensions of products using specialized tools like calipers, micrometers, and gauges to ensure they meet specified tolerances.

Example – In a machine parts factory, each component is checked for proper dimensions to ensure it fits correctly in assembly.

4 **Performance Testing** – Conducting tests to ensure that products function as intended under specific conditions.

Example: In an electrical appliance manufacturing facility, each product undergoes performance testing, such as running washing machines through full cycles to verify operation.

5 **Non-Destructive Testing (NDT)** – Using methods that do not damage the product to assess its properties, such as ultrasonic, radiographic, or magnetic particle inspections.

Example – In the aerospace industry, aircraft components undergo NDT to detect internal defects without affecting structural integrity.

6 **Documentation and Record-Keeping** – Recording the results of inspections, tests, and analysis for traceability, accountability, and future reference.

Example – A food manufacturing plant maintains detailed records of each QC inspection, including batch numbers, test results, and the date of inspection.

10.2.6.17 Types of Quality Control and Inspection Areas

1 **Quality Assurance (QA) Laboratory** – A controlled environment where detailed chemical, physical, and performance tests are conducted on product samples.

Example – In a cosmetics manufacturing company, a QA lab tests the stability, safety, and efficacy of skincare products to meet health and safety regulations.

2 **In-Line Quality Control Station** – Integrated QC stations directly on the production line to inspect products at various stages of manufacturing.

Example – In an automotive factory, in-line QC stations check the assembly of engine components before the next stage in the line.

3 **Final Inspection Area** – The last QC station before products are shipped or stored. This area verifies that products are defect-free and meet all quality criteria.

Example – In a furniture factory, the final inspection area checks that all assembled furniture is stable, aesthetically pleasing, and free from defects.

4 **Receiving Inspection Area** – A QC area that inspects incoming raw materials and components before they are used in production.

Example – In an electronics assembly plant, receiving inspection checks that purchased components like chips, capacitors, and resistors meet specifications.

10.2.6.18 Key Considerations in Designing the Quality Control and Inspection Area

1 **Space and Layout** – The QC area should have sufficient space for testing equipment, inspection tools, and safe handling of products. A well-organized layout supports efficient workflow and minimizes errors.

Example – A pharmaceutical QC lab is designed with separate areas for sample collection, testing, and documentation to avoid cross-contamination.

2 **Lighting and Visibility** – Proper lighting is essential to detect defects accurately during visual inspections. Good lighting reduces the likelihood of oversight.

Example – In a jewelry manufacturing plant, bright lighting is used to inspect gemstones for clarity, inclusions, and flaws.

3 **Environmental Control** – Some QC areas require controlled environments for accurate testing, including temperature and humidity control.

Example – In a semiconductor manufacturing facility, the QC area is a clean room environment to prevent contamination during inspection.

4 **Access and Security** – The QC area should be accessible only to authorized personnel to maintain product integrity and confidentiality.

Example – In a defense equipment manufacturing plant, the QC lab is secured with restricted access to protect sensitive information.

10.2.6.19 Challenges in Quality Control and Inspection

- **Human Error** – Manual inspections may result in errors due to fatigue or oversight, highlighting the importance of automation in repetitive checks.
- **Equipment Calibration** – QC tools and equipment require regular calibration to maintain accuracy, making ongoing maintenance essential.
- **Time Constraints** – Performing thorough inspections can be time-consuming, potentially slowing down production if not managed effectively.

10.2.6.20 Examples of Quality Control and Inspection in Different Industries

1 **Food Industry** – A dairy processing plant uses microbiological testing to verify the safety of milk products and ensures packaging integrity through visual inspections.

2 **Automotive Industry** – Car manufacturers perform crash tests and road simulations in the QC area to ensure vehicle safety before mass production.

3 **Textile Industry** – Fabric quality is assessed using color consistency tests, tear resistance measurements, and seam strength checks.

4 **Electronics Industry** – QC in electronics involves electrical testing for circuit boards, visual checks for soldering quality, and thermal testing for durability.

5 **Construction Industry** – In a construction project, concrete samples are taken to QC labs to test for compressive strength, ensuring the materials meet safety standards.

10.2.6.21 Importance of the Quality Control and Inspection Area

- **Customer Satisfaction** – By ensuring consistent product quality, QC helps maintain customer trust and satisfaction.
- **Cost Reduction** – Identifying defects early in production minimizes rework costs, waste, and product recalls.
- **Brand Reputation** – A well-maintained QC system contributes to a strong brand reputation, signaling reliability and attention to detail.

- **Regulatory Compliance** – Proper QC ensures that products meet legal and industry standards, avoiding penalties and fines.

10.2.6.21.1 Conclusion

The Quality Control and Inspection Area (Figure 10.19) is a critical component in any manufacturing or production setup, ensuring that products meet quality standards and specifications. By employing a combination of manual, automated, and high-tech methods, the QC area acts as the last defense in safeguarding product integrity, reducing costs, and fostering customer satisfaction. Properly organized and equipped QC areas contribute to the overall success and efficiency of any industrial or commercial operation.

Figure 10.19: Quality Control and Inspection Area

10.2.6.22 Layout Optimization Considerations

Layout Optimization involves designing or improving the physical arrangement of equipment, workstations, and facilities within a manufacturing plant, warehouse, or office to maximize efficiency, productivity, safety, and space utilization. The goal is to create a workflow that reduces wasted time, resources, and motion, while facilitating the smooth movement of materials, information, and people.

Below are key considerations for layout optimization with relevant examples:

1 Material Flow

Ensuring that the flow of materials throughout the facility is logical and uninterrupted is crucial for layout optimization. Materials should move seamlessly from one processing area to another to minimize delays.

Example – In a **car manufacturing plant**, materials flow from the raw materials storage to the stamping, assembly, paint, and final inspection sections. The layout is designed to minimize the distance each part travels, reducing transportation time.

2 Space Utilization

Space is a valuable asset in any facility. Optimizing layout involves making the best use of available space without overcrowding, allowing for easy movement and flexibility.

Example – In a **warehouse**, vertical shelving is used to store products, maximizing the use of vertical space and freeing up floor space for movement and operations.

3 Flexibility

A flexible layout is adaptable to changes in production volume, product types, or processes without requiring significant modifications. Flexibility is essential for industries that produce various products or experience seasonal fluctuations.

Example – A **cellular manufacturing layout** in an electronics factory allows for quick reconfiguration of workstations to accommodate different product models without major disruptions.

4 Minimizing Travel Distance

Reducing the travel distance for workers, machines, and materials minimizes waste and improves efficiency. A well-optimized layout aims to shorten the distance covered during the production process.

Example – In a **food processing plant**, the cooking, packaging, and storage areas are arranged in close proximity to reduce the time and effort needed to transfer products between stages.

5 Safety and Ergonomics

Safety is a priority in layout optimization. The layout should minimize hazards, ensure safe distances between equipment, and provide clear paths for emergency exits. Ergonomically designed workstations reduce worker fatigue and injury.

Example – In a **chemical plant**, hazardous materials are stored in a separate, ventilated area away from the production line to reduce the risk of accidents. Safety barriers and marked walkways direct traffic flow to avoid collisions.

6 Workflow Efficiency

The layout should be designed to promote smooth workflow by minimizing bottlenecks, reducing waiting times, and ensuring that processes are in a logical sequence.

Example – In a **pharmaceutical factory**, the workflow is arranged in a linear sequence: from raw material preparation to mixing, tableting, packaging, and quality control. This setup eliminates bottlenecks and ensures a streamlined production line.

7 Integration of Technology

Modern manufacturing and office environments often rely on automation and technology. Optimizing layout includes integrating machinery, robotics, and digital systems in a way that enhances productivity.

Example – In an **automated packaging facility**, robotic arms are strategically positioned along the conveyor belt to perform packaging, sealing, and labeling tasks efficiently. The layout is adjusted to accommodate the robots' range of motion.

8 Handling of Equipment and Machinery

Heavy machinery or equipment should be positioned in a way that facilitates easy access for maintenance, repairs, and replacement. Large equipment should not obstruct workflow or hinder movement.

Example – In a **metal fabrication workshop**, welding machines are placed along the perimeter of the shop floor, with enough space for forklifts to move large metal sheets and completed parts.

9 Storage and Inventory Management

Storage areas should be conveniently located to minimize the handling time for raw materials and finished goods. Efficient inventory management can significantly reduce operational costs.

Example – In a **retail distribution center**, high-demand items are stored near the loading dock to reduce the time needed for picking and shipping. Less frequently accessed inventory is stored further away.

10 Communication and Coordination

The layout should facilitate easy communication and coordination among employees, departments, and teams. Clear sightlines and proximity can improve collaboration.

Example – In an **open-plan office**, departments that frequently collaborate are placed close to each other, and there are no physical barriers to hinder interaction. Conference rooms are centrally located for easy access.

11 Environmental Considerations

Optimizing layout includes environmental factors such as lighting, ventilation, and temperature control, which can affect worker productivity and comfort.

Example – In a **green building**, windows are strategically placed to maximize natural lighting, reducing the need for artificial light. Ventilation systems are designed to improve airflow and maintain a comfortable temperature.

12 Cost Efficiency

The cost of layout changes should be justified by the benefits gained in terms of efficiency, productivity, or quality. Avoid expensive changes unless they provide a clear return on investment.

Example – In a **small manufacturing unit**, relocating equipment to create a more linear workflow may improve efficiency without the need for additional machinery, justifying the cost of rearranging the layout.

10.2.6.23 Types of Layouts for Optimization

Below are some examples of layout types that are commonly optimized for different scenarios:

1 Process Layout

Designed for industries with low-volume, high-variety production. Similar processes are grouped together.

Example – In a **job shop**, where custom products are produced, machinery is arranged based on function, like cutting, welding, and assembly sections.

2 Product Layout

Best suited for high-volume, low-variety production. The layout is linear, following the sequence of production.

Example – An **automobile assembly line**, where the product moves from one workstation to another in a defined order.

3 Cellular Layout

Uses a combination of process and product layout, grouping machines for similar products into cells.

Example – In a **furniture factory**, a cell may be set up for producing a specific chair model, combining cutting, sanding, and assembly in one area.

4 Fixed-Position Layout

The product remains in one place while workers and machinery come to it.

Example – In **shipbuilding**, the vessel remains stationary, and the workforce, tools, and parts are brought to the ship.

10.2.6.24 Conclusion

Optimizing layout is a fundamental part of designing an efficient and effective workspace, be it a manufacturing plant, warehouse, or office. By considering factors like material flow, space utilization, flexibility, safety, and cost, companies can create environments that support their operational goals. The result is a smoother workflow, enhanced productivity, better use of space, and improved safety standards—all contributing to a more competitive and profitable organization.

10.2.6.25 Do Work Visits to Industries – Industry Visit

"Do work visits to industries," also known as industry visits or field visits, involve touring a business or manufacturing plant to observe operations, understand processes, and gather insights directly from working environments. These visits are often organized for students, trainees, or employees to bridge theoretical knowledge with practical, real-world applications, enhance learning, and support professional development. Industry visits can vary depending on the industry and objectives, from observing production lines to engaging with management about operational challenges and best practices.

10.2.6.26 Purpose and Benefits of Industry Visits

1 **Learning Practical Applications** – Industry visits allow participants to see how theoretical knowledge applies practically. For instance, engineering students studying assembly lines may visit an automotive manufacturing plant to observe how different components are assembled into a complete vehicle.

2 **Enhancing Knowledge and Skills** – Exposure to current technologies and methodologies can expand participants' technical knowledge and skillsets.

3 **Building Professional Networks** – Industry visits often facilitate networking opportunities with professionals and industry experts.

4 **Career Exposure** – Such visits help participants explore different roles and career paths within an industry, aiding career planning.

10.2.6.27 Examples of Industry Visits by Sector

1 Manufacturing Sector

Example – Engineering students visiting a car manufacturing plant like Toyota or Ford could observe automated welding robots, conveyer belt systems, and quality control measures in action.

Objective – Learn about manufacturing operations, lean manufacturing practices, and automated systems. Students also get an opportunity to understand production scales, quality assurance, and safety protocols.

2 Pharmaceutical Sector

Example – Pharmacy students touring a pharmaceutical manufacturing facility like Pfizer or GSK to understand how medicines are manufactured, quality-tested, and packaged.

Objective – Gain insights into Good Manufacturing Practices (GMP), drug formulation, and regulatory compliance. They may observe quality control labs and sterile environments, learning the strict controls needed in pharmaceutical production.

3 Construction Industry

Example – Architecture and civil engineering students visiting a construction site, such as a high-rise building or infrastructure project.

Objective – See the application of architectural plans, project management practices, and safety procedures on a live construction site. Observing the sequence of construction, materials used, and interaction among teams can provide insight into the complexities of managing a construction project.

4 Information Technology Sector

Example – IT students visiting a data center managed by tech companies like Google or Amazon.

Objective: Understand how data storage and cloud infrastructure are organized, secured, and maintained. Students can learn about server maintenance, cybersecurity measures, and data backup systems essential for managing high volumes of data.

5 Energy and Utilities Sector

Example – Students visiting a renewable energy plant, such as a solar farm or wind power station.

Objective: Learn about sustainable energy generation, storage, and distribution. Observing energy generation operations can help students understand the technical and environmental considerations in renewable energy industries.

6 Hospitality and Service Sector

Example – Hospitality management students visiting a hotel chain to observe its operations.

Objective: Gain insight into front-desk management, room service, housekeeping, event coordination, and customer service. This allows students to learn about service quality standards, customer engagement strategies, and workflow management.

7 Agriculture and Food Processing Sector

Example – Agricultural students visiting a dairy farm or a food processing facility like Nestle.

Objective: Learn about sustainable farming practices, food safety standards, processing techniques, and packaging. Participants can understand how farm products are processed into consumer goods and the technology involved in maintaining food quality and safety.

10.2.6.28 Organizing Effective Work Visits

To maximize the effectiveness of industry visits, companies and educators often follow these practices:

- **Pre-visit Preparation**: Participants are briefed on the objectives, the company, and its operations, allowing them to prepare questions and identify areas of interest.
- **Guided Tours** – An on-site professional guide the visit, explaining processes and answering questions. This allows for interactive learning and clarifications.
- **Hands-on Experience** – Where possible, participants may be allowed to engage in safe, supervised activities to get a feel for certain tasks, such as assembling parts on a production line.
- **Post-visit Review** – Participants often share feedback, discuss observations, and reflect on learning outcomes after the visit to reinforce the knowledge gained.

10.2.2.29 Summary

In summary, work visits to industries are valuable learning experiences that provide a deeper understanding of real-world operations, broaden participants' perspectives, and help them relate academic theories to actual industrial practices.

CHAPTER ELEVEN

11.0 SYSTEMATIC FAULT DIAGNOSIS: PRINCIPLES, TECHNIQUES, AND ANALYTICAL TOOLS FOR INDUSTRIAL MAINTENANCE

11.1 LOGICAL METHODS AND MAINTENANCE PROBLEMS IN INDUSTRY

Maintaining equipment, machinery, and infrastructure in the industry requires strategic approaches to ensure reliability, efficiency, and safety. Effective **maintenance management** involves the use of logical methods, tools, and processes to identify and solve maintenance problems proactively.

11.1.1 Logical Methods in Maintenance

11.1.1.1 Preventive Maintenance (PM)

- **Definition** – Preventive maintenance is a proactive approach where regular inspections, servicing, and repairs are scheduled based on a time interval or usage metrics.
- **Method** – Equipment is inspected and serviced at predetermined intervals to prevent unexpected failures.
- **Advantages** – Reduces downtime, extends equipment life, and lowers overall repair costs.
- **Challenges** – Requires resources and scheduling, which can lead to over-maintenance if intervals are not accurately defined.

11.1.1.2 Predictive Maintenance (PdM)

- **Definition** – Predictive maintenance uses data, sensors, and monitoring tools to predict when equipment is likely to fail. Maintenance is carried out just before a failure is predicted to occur.
- **Method** – Analysis of vibration, temperature, noise, and other data to assess equipment health.
- **Advantages** – Minimizes unplanned downtime and optimizes maintenance schedules.
- **Challenges** – Requires investment in monitoring technology and skilled personnel to interpret data.

11.1.1.3 Condition-Based Maintenance (CBM)

- **Definition** – Maintenance tasks are performed when monitoring equipment indicates that a threshold has been reached, signaling potential failure.
- **Method** – Using sensors and diagnostics to monitor parameters like temperature, pressure, and oil quality.
- **Advantages** – Maintenance is carried out only when necessary, saving time and costs.
- **Challenges** – Can be costly to implement and requires expertise in interpreting condition indicators.

11.1.1.4 Reliability-Centered Maintenance (RCM)

- **Definition** – RCM is a systematic approach to determine the maintenance needs of a physical asset to ensure it performs its intended function.
- **Method** – Analysing failures, critical equipment, and the effects of equipment failure to develop a tailored maintenance strategy.
- **Advantages** – Focuses on reliability, safety, and performance while reducing unnecessary maintenance.
- **Challenges**: In-depth analysis is time-consuming and requires expert knowledge.

11.1.1.5 Root Cause Analysis (RCA)

- **Definition** – RCA is a problem-solving technique used to identify the underlying cause of equipment failures.
- **Method** – Investigating failures, analysing data, and identifying patterns to prevent recurrence.
- **Advantages** – Reduces recurrence of problems and increases the reliability of equipment.
- **Challenges** – Requires time and effort to analyse data and identify the true cause.

11.1.1.6 Total Productive Maintenance (TPM)

- **Definition** – TPM is an all-inclusive maintenance strategy that involves every member of the organization, from operators to management.
- **Method** – Encouraging autonomous maintenance by operators, regular training, and cross-functional teams.
- **Advantages** – Improves overall equipment effectiveness (OEE) and engages employees.

- **Challenges** – Involves cultural change, training, and commitment at all organizational levels.

11.1.2 Common Maintenance Problems

11.1.2.1 Lack of Proper Planning and Scheduling

- **Problem** – Maintenance work is often reactive rather than planned, leading to last-minute fixes and rushed repairs.
- **Solution** – Implement a detailed preventive or predictive maintenance plan with a clear schedule for inspections and repairs.
- **Impact** – Proper planning reduces unplanned downtime, prevents unexpected failures, and allows efficient resource allocation.

11.1.2.2 Equipment Downtime and Unexpected Failures

- **Problem** – Equipment can fail suddenly, causing operational delays and increased costs.
- **Solution** – Use predictive and condition-based maintenance techniques to monitor equipment health and predict failures.
- **Impact** – Preventing unexpected downtime saves costs, ensures steady production, and avoids emergency repairs.

11.1.2.3 Insufficient Maintenance Resources

- **Problem** – A shortage of skilled personnel, tools, or replacement parts can hinder effective maintenance.
- **Solution** – Invest in training, ensure inventory management for spare parts, and use computerized maintenance management systems (CMMS) to track resources.
- **Impact** – Skilled personnel and well-managed resources lead to maintenance that the operations are more effective.

11.1.2.4 Inadequate Documentation and Record Keeping

- **Problem** – Without proper documentation, tracking maintenance history, service intervals, and recurring problems is difficult.
- **Solution** – Use digital tools like CMMS to maintain detailed records of maintenance activities, inspections, and equipment performance.

- **Impact** – Accurate records help predict failures, identify patterns, and improve decision-making.

11.1.2.5 Incorrect Diagnosis of Problems

- **Problem** – Misdiagnosing a fault can lead to incorrect repairs, increased costs, and further damage to equipment.
- **Solution** – Conduct thorough root cause analysis (RCA) and involve experienced personnel in the diagnosis process.
- **Impact** – Accurate diagnosis reduces repair time, costs, and recurrence of the problem.

11.1.2.6 Over-Maintenance and Unnecessary Repairs

- **Problem** – Performing too much maintenance can increase costs, wear out components, and cause unnecessary downtime.
- **Solution** – Use predictive or condition-based maintenance strategies to focus efforts where needed.
- **Impact** – Reducing over-maintenance leads to lower costs, increased productivity, and extended equipment lifespan.

11.1.2.7 Human Error in Maintenance

- **Problem** – Mistakes made by maintenance personnel during inspections, repairs, or installations can lead to equipment failures.
- **Solution** – Provide regular training, implement standardized procedures, and encourage a culture of safety and accountability.
- **Impact** – Reducing human error improves maintenance quality and equipment reliability.

11.1.2.8 Inconsistent or Inadequate Maintenance Standards

- **Problem** – Inconsistent standards and procedures can result in subpar maintenance practices and poor equipment performance.
- **Solution** – Develop standard operating procedures (SOPs) and adhere to industry best practices.
- **Impact** – Consistency in maintenance improves overall equipment effectiveness and reliability.

11.1.3 Addressing Maintenance Problems

11.1.3.1 Implementing Technology

- Use computerized maintenance management systems (CMMS) to streamline maintenance tasks, schedule repairs, track equipment history, and manage resources.
- Integrate **Industrial Internet of Things (IIoT)** sensors for real-time monitoring of equipment conditions.

11.1.3.2 Training and Upskilling

- Invest in continuous training for maintenance personnel to keep them updated on the latest tools, technology, and best practices.
- Encourage cross training to ensure maintenance teams are versatile and can handle diverse challenges.

11.1.3.3 Effective Communication

- Establish clear communication between maintenance teams, operators, and management to ensure that issues are reported accurately and promptly.
- Use maintenance dashboards to provide visibility into ongoing tasks, pending inspections, and resource allocation.

11.1.3.4 Use of Maintenance Metrics

- Track maintenance performance using key metrics such as **Mean Time Between Failures (MTBF)**, **Mean Time to Repair (MTTR)**, and equipment downtime.
- Analyse these metrics regularly to identify trends and adjust maintenance strategies accordingly.

11.1.3.5 Continuous Improvement

- Encourage a culture of continuous improvement in maintenance practices by learning from failures and near misses.
- Regularly update maintenance strategies based on feedback, technological advancements, and industry standards.

11.1.4 Summary

In summary, maintenance in industry requires a combination of logical methods, strategic planning, and consistent effort to prevent breakdowns, reduce costs, and maintain safety. By implementing effective maintenance methods, addressing common problems, and investing in training and technology, companies can enhance the reliability and efficiency of their equipment and infrastructure, leading to better overall performance and safety.

11.2 PRINCIPLES OF FAULT ANALYSIS AND DIAGNOSIS

Fault analysis and diagnosis are essential in identifying the root causes of equipment failures, system errors, or performance issues in industrial and technical systems. These processes aim to prevent further damage, improve reliability, and enhance operational efficiency.

11.2.1 Key Principles and Examples of Fault Analysis

11.2.1.1 Data Collection

- **Principle** – The first step in fault analysis is to gather data related to the fault. This includes collecting information about operating conditions, fault symptoms, equipment parameters, and environmental factors. The goal is to capture any anomalies or deviations that might indicate a problem.
- **Example** – In a manufacturing plant, if a machine starts vibrating unusually, sensors might be used to record the vibration levels, temperature, pressure, and sound to understand the fault's context.

11.2.1.2 Symptom Identification

- **Principle** – Clearly define and identify the symptoms of the fault. These symptoms are observable indications of an underlying issue and may include unusual noises, vibrations, error messages, performance drops, or visual defects.
- **Example** – If a conveyor belt is moving slower than usual, the symptoms could be loud screeching sounds, slow speed, or irregular movement patterns. These observations will guide the diagnosis.

11.2.1.3 Fault Isolation

- **Principle** – Fault isolation is the process of narrowing down the potential causes of a problem. This involves ruling out unrelated components and systems to focus on the area where the issue is likely to exist. It often includes the use of diagnostic tools and tests.

- **Example** – In an electrical circuit, if a light bulb is not turning on, fault isolation might involve checking the power supply, switch, wiring, and bulb individually to identify which part is causing the problem.

11.2.1.4 Root Cause Analysis (RCA)

- **Principle** – RCA is a systematic method to determine the primary cause of a fault. It goes beyond the symptoms to find the underlying problem. Techniques like 5 Whys, Fishbone Diagrams, or Failure Mode and Effects Analysis (FMEA) are often used to get to the root cause.
- **Example** – In a factory, if a motor overheats and stops, RCA might involve asking a series of questions: *Why did the motor overheat?* (Blocked airflow) → *Why was the airflow blocked?* (Filter was clogged) → *Why was the filter clogged?* (Regular maintenance was skipped).

11.2.1.5 Hypothesis Formulation

- **Principle** – Based on the data and observations, formulate hypotheses regarding the possible causes of the fault. Each hypothesis should be testable and based on evidence gathered during data collection.
- **Example** – If a car is not starting, potential hypotheses might include a dead battery, faulty starter, or fuel system issue. These hypotheses guide the testing process.

11.2.1.6 Testing and Verification

- **Principle** – Conduct tests to verify or refute the formulated hypotheses. This step involves applying diagnostic tests, simulations, or inspections to confirm the presence of a fault in a specific component or system.
- **Example** – To verify if a car's battery is dead, one could use a multimeter to measure the battery voltage. If the voltage is below the acceptable range, the hypothesis is confirmed.

11.2.1.7 Fault Diagnosis

- **Principle** – Once the root cause has been identified, diagnose the fault by determining what component or system failed and why. Diagnosis may involve analysing failure mechanisms, material degradation, or operational errors.
- **Example** – In a computer system, if an application crashes frequently, diagnostics might reveal a software bug due to a specific update, leading to a more targeted solution.

11.2.1.8 Corrective Action

- **Principle** – Implement corrective actions to fix the identified fault. This may involve repairing, replacing, or recalibrating components, updating software, or modifying processes to prevent recurrence.
- **Example** – If a pump fails due to worn-out bearings, replacing the bearings and scheduling regular lubrication can serve as corrective actions to prevent future failures.

11.2.1.9 Verification of Solution

- **Principle** – After corrective action has been taken, verify the solution's effectiveness by monitoring the system to ensure that the fault does not recur. This may involve retesting or observing the equipment's performance over time.
- **Example** – After replacing the bearings in a pump, monitor the pump's vibration levels and temperature to ensure it operates smoothly without repeating the failure.

11.2.1.10 Documentation and Reporting

- **Principle** – Document the fault analysis process, findings, actions taken, and lessons learned. This information can be used for training, future reference, and continuous improvement.
- **Example** – In an industrial plant, if a hydraulic system fails due to a leaking hose, the incident, analysis, and corrective action should be documented in the maintenance log for future technicians.

11.2.2 Examples of Fault Analysis and Diagnosis

11.2.2.1 Electrical Circuit Fault

- **Problem** – A circuit breaker in a building trips frequently.
- **Step-by-Step Analysis**

1 **Data Collection** – Check the breaker specifications, load conditions, and environmental factors.

2 **Symptom Identification** – Record when the breaker trips (e.g., when multiple devices are in use).

3 **Fault Isolation** – Disconnect devices one by one to see which device or combination causes the trip.

4 **RCA** – Determine if the breaker trips due to overload, short circuit, or faulty wiring.

5 **Hypothesis** – Hypothesize that an overloaded circuit or a malfunctioning appliance is causing the trips.

6 **Testing** – Use a clamp meter to measure current load and compare it with the breaker's capacity.

7 **Diagnosis** – If the load exceeds capacity, it is an overload issue; if not, investigate other potential faults.

8 **Corrective Action** – Rebalance the load across circuits or replace the faulty appliance.

9 **Verification** – Monitor the circuit with the new load distribution.

10 **Documentation** – Record the incident, analysis, and resolution for future reference.

11.2.2.2 Machinery Vibration Issue

- **Problem** – A rotating machine exhibits unusual vibrations.
- **Step-by-Step Analysis**

1 **Data Collection** – Collect vibration data using sensors, noting frequency, amplitude, and direction.

2 **Symptom Identification** – Identify changes in vibration patterns compared to normal operation.

3 **Fault Isolation** – Isolate potential causes such as misalignment, unbalance, or bearing wear.

4 **RCA** – Determine whether the issue is due to wear, imbalance, or external factors like loose components.

5 **Hypothesis** – Misalignment could be the primary cause.

6 **Testing** – Use a laser alignment tool to check the alignment of the machine.

7 **Diagnosis** – If misaligned, note which components are out of tolerance.

8 **Corrective Action** – Realign the machine components according to manufacturer specifications.

9 **Verification** – Monitor the machine after alignment to ensure vibrations return to acceptable levels.

10 **Documentation** – Record the incident, findings, and alignment process.

11.2.2.3 Computer Network Fault

- **Problem** – A network experiences intermittent connection drops.

- **Step-by-Step Analysis**

1 **Data Collection** – Gather logs of connection events, network load, and error messages.

2 **Symptom Identification** – Identify specific times, devices, or applications associated with the drops.

3 **Fault Isolation** – Test the network components, including routers, switches, and cables.

4 **RCA** – Use packet-sniffing tools to detect if a specific device or network segment is causing interference.

5 **Hypothesis** – Suspect that an overloaded router or faulty cable is causing disruptions.

6 **Testing** – Replace the suspected cable or increase router capacity and monitor network stability.

7 **Diagnosis** – Confirm whether the cable was faulty or if increased capacity resolves the issue.

8 **Corrective Action** – Replace faulty hardware or optimize network load distribution.

9 **Verification** – Test the network under typical load conditions to confirm stability.

10 **Documentation** – Log the diagnostic steps, root cause, and changes made to the network.

11.2.3 Summary

In summary, fault analysis and diagnosis involve a systematic approach to identifying and solving issues in technical and industrial systems. By following logical methods—data collection, symptom identification, hypothesis testing, and corrective action—industries can minimize downtime, improve equipment reliability, and optimize performance. Documenting each step provides valuable information for future fault management, leading to a proactive and efficient maintenance culture.

11.3 IDENTIFYING AND LOCATING FAULTS: METHODS AND EXAMPLES

In any industrial or technical setting, identifying and locating faults is crucial for maintaining equipment reliability, minimizing downtime, and ensuring safety. Fault identification involves detecting that a problem exists, while fault location pinpoints the exact source of the issue. Effective fault management relies on systematic methods and tools to diagnose and address faults quickly and accurately.

11.3.1 Methods for Identifying and Locating Faults

11.3.1.1 Visual Inspection

- **Principle** – Conduct a physical examination of equipment or systems to detect visible signs of damage or abnormalities, such as cracks, leaks, corrosion, loose components, or wear.
- **Application** – Often the first step in fault detection, visual inspection can reveal obvious defects.
- **Example** – In a hydraulic system, a technician might visually inspect hoses and fittings for leaks or damage if a drop in pressure is observed.

11.3.1.2 Monitoring and Measurement

- **Principle** – Use instruments to monitor and measure operational parameters like temperature, pressure, voltage, current, vibration, or noise to identify deviations from normal values.
- **Application** – Monitoring is effective for detecting changes in performance that may indicate a fault.
- **Example** – In a motor, measuring vibration levels with an accelerometer can help detect imbalance or bearing wear before failure occurs.

11.3.1.3 Diagnostic Tools and Test Equipment

- **Principle** – Employ diagnostic tools and equipment to conduct targeted tests on systems and components to identify faults. Common tools include multimeters, oscilloscopes, thermal cameras, or vibration analysers.
- **Application** – Useful for electrical, mechanical, and electronic systems where precision is needed.
- **Example** – An electrician might use a multimeter to check for voltage drops in an electrical circuit, helping to identify a faulty connection.

11.3.1.4 Simulation and Fault Replication

- **Principle** – Simulate normal and abnormal conditions or replicate faults in a controlled environment to observe system behavior and identify the source of the problem.
- **Application** – Commonly used for complex systems like software or automation systems, where replicating a fault can help trace its origin.

- **Example** – In a software-controlled machine, a technician might use a simulation tool to recreate an error scenario to identify if a software bug or hardware issue causes it.

11.3.1.5 Signal Tracing and Testing

- **Principle** – Trace signals (electrical, mechanical, or data) step-by-step through a system to identify the point at which the signal becomes abnormal.
- **Application** – Effective for electrical circuits, data communication networks, or process control systems.
- **Example** – In an electrical circuit, a technician may use a logic probe to trace signals from input to output, identifying the point where the signal fails.

11.3.1.6 Component Substitution and Bypass Testing

- **Principle** – Substitute suspect components with known good ones or temporarily bypass a component to see if the fault is eliminated.
- **Application** – Common for electronic or electrical systems where component failure is suspected.
- **Example** – If a printer's power supply is suspected to be faulty, swapping it with a working unit can confirm if the power supply is the issue.

11.3.1.7 Error Code Analysis and Diagnostic Software

- **Principle** – Many modern systems include diagnostic software that provides error codes or logs, indicating the potential fault location.
- **Application** – Often used in automotive, manufacturing, or IT systems with built-in diagnostics.
- **Example** – A car's on-board diagnostic (OBD) system might provide an error code indicating a faulty oxygen sensor, guiding technicians to check and replace the sensor.

11.3.1.8 Historical Data and Trend Analysis

- **Principle** – Analyse historical performance data, maintenance records, and trends to identify recurring faults and predict potential failures.
- **Application** – Effective for equipment with long-term data history, allowing proactive maintenance.

- **Example** – In a manufacturing plant, analysing past data on machine breakdowns can reveal patterns, such as increased failure rates after certain maintenance intervals, helping to adjust the maintenance schedule.

11.3.2 Examples of Fault Identification and Location

11.3.2.1 Identifying a Hydraulic Leak

- ***Scenario*** – A hydraulic press in a factory shows a gradual decline in pressure during operation.

Identification

- ***Visual Inspection*** – Check for visible leaks around hydraulic lines, cylinders, and valves.
- ***Measurement*** – Use a pressure gauge to monitor pressure drops in different sections of the hydraulic system.

Fault Location

- Isolate different sections of the hydraulic circuit by closing valves and testing each segment separately.
- The leak is identified in a specific section of the circuit, pinpointing a worn seal in a cylinder.

11.3.2.2 Electrical Circuit Short Circuit

- ***Scenario*** – A lighting circuit in a building trips the circuit breaker repeatedly.

  Identification

- **Initial Check** – Reset the breaker and observe if the circuit trips again. Use a multimeter to measure current in the circuit.
- ***Symptom Observation*** – Note when and under what conditions the circuit breaker trips (e.g., when a specific light switch is turned on).

Fault Location

- Use a multimeter to test continuity in different sections of the circuit.
- Trace wiring to locate a pinched or exposed wire that is short-circuiting when the switch is activated.

11.3.2.3 Computer Network Connectivity Issue

- ***Scenario*** – Users report intermittent internet connection drops in an office network.

Identification

- ***Monitoring*** – Use network monitoring tools to track packet loss, latency, and bandwidth utilization.
- ***Observation*** – Check if the problem affects all users or only specific segments of the network.

Fault Location

- Use a cable tester to check Ethernet cables.
- Ping each network switch to identify if one specific switch is causing the issue. The problematic switch is replaced, restoring normal connectivity.

11.3.2.4 Overheating Motor in Manufacturing

- ***Scenario*** A motor in a manufacturing line is overheating and tripping thermal protection.

Identification

- ***Data Collection*** – Use a thermal camera to identify hot spots on the motor and measure the motor's operating temperature.
- ***Symptom Monitoring*** – Check if the motor overheats at a particular load or operational speed.

Fault Location

- Use a tachometer to measure the motor's speed and ensure it matches specifications.
- Check the motor's cooling fan and ventilation system for obstructions. A blocked cooling duct is found to be the issue.

11.3.2.5 Software Bug in Industrial Control System

- ***Scenario*** – An automated conveyor system occasionally stops unexpectedly, disrupting production.

Identification

- ***Error Analysis*** – Use the system's diagnostic software to check for error codes when the fault occurs.
- ***Data Collection*** – Review error logs to identify patterns in the shutdowns.

Fault Location

- Replicate the fault by simulating similar load and speed conditions to observe system behavior.
- Use debugging tools to trace the code path, revealing a software bug in the speed control algorithm that triggers under specific load conditions.

11.3.2.6 Mechanical Fault in Gearbox

- ***Scenario*** – A gearbox in a conveyor system produces a grinding noise and reduced output power.

Identification

- ***Vibration Analysis*** – Use a vibration analyser to detect unusual patterns that might indicate bearing or gear damage.
- ***Oil Analysis*** – Inspect the gearbox oil for metal particles that indicate internal wear.

Fault Location

- Disassemble the gearbox and inspect individual gears and bearings for wear or misalignment.
- A damaged gear is identified and replaced, eliminating the grinding noise.

11.3.3 Summary

In summary, identifying and locating faults require a structured approach using a combination of tools, observation, and analysis. By following a systematic process—data collection, monitoring, testing, and diagnosis—technicians and engineers can detect faults early, minimize downtime, and maintain the efficiency and reliability of industrial systems. The right diagnostic methods, combined with skilled observation and technical knowledge, lead to quicker fault resolution and a safer working environment.

11.4 USES OF FLOW DIAGRAMS, ALGORITHMIC ANALYSIS, AND DEPENDENCY CHARTS

Flow diagrams, algorithmic analysis, and dependency charts are valuable tools in system design, troubleshooting, project management, and process optimization. Each of these tools provides a different view of the process or system being analysed, helping identify dependencies, streamline operations, and optimize performance. Below is an overview of each with examples:

11.4.1 Flow Diagrams

- **Definition** – Flow diagrams (or flowcharts) are graphical representations of a process, workflow, or system. They use standardized symbols, such as arrows and shapes, to illustrate the sequence of actions, decisions, and steps within a process.

Uses

- ***Process Visualization*** – Makes complex processes easy to understand by mapping them out step-by-step.
- ***Troubleshooting*** – Identifies bottlenecks or inefficiencies by visualizing each step of a process.
- ***Training*** – Provides an intuitive, visual guide for new team members or trainees.
- ***Documentation*** – Serves as a formal documentation of standard operating procedures (SOPs).
- **Example** – In a **Manufacturing Process**, a flow diagram could illustrate the steps from material sourcing to final product assembly. It might start with "Material Receipt," followed by "Quality Inspection," "Assembly," "Testing," and end at "Packaging." Decision points could include a quality check after testing, where the process either continues to packaging or returns to rework if defects are found.

11.4.2 Algorithmic Analysis

- **Definition** – Algorithmic analysis is the process of evaluating an algorithm to determine its efficiency, performance, and scalability. It typically focuses on time complexity (how execution time increases with input size) and space complexity (memory usage).
- **Uses**
- ***Performance Optimization*** – Helps optimize code or processes by selecting the most efficient algorithm.
- ***Scalability Analysis:*** Determines how an algorithm performs as data size increases, crucial for systems expected to handle large volumes of data.
- ***Cost-Efficiency*** – Ensures efficient use of computational resources, reducing operational costs.
- ***Decision-Making*** – Assists in selecting the best algorithm based on application requirements and constraints.
- ***Example*** – Suppose a company needs to sort a large dataset of customer records. Comparing algorithms like **Quick Sort**, **Merge Sort**, and **Bubble Sort** through

algorithmic analysis can reveal that Quick Sort and Merge Sort have better time complexities (O(nlog n)) than Bubble Sort (O(n^2)), making them better choices for large datasets.

11.4.3 Dependency Charts

- **Definition** – Dependency charts map out the dependencies between tasks, components, or systems within a larger project or system. They display the relationships and hierarchies that define the order in which tasks or components must be completed.

Uses

- ***Project Management*** – Helps organize tasks based on dependencies, facilitating scheduling and resource allocation.
- ***Risk Assessment*** – Identifies critical dependencies where delays or issues could affect the entire project.
- ***Resource Optimization*** – Allocates resources efficiently by understanding which tasks can run in parallel and which must be completed sequentially.
- ***Troubleshooting and Maintenance*** – Provides insights into potential impacts of a failure in one part of a system on dependent components.
- ***Example*** – In **Software Development**, a dependency chart could show that coding tasks depend on design tasks, which in turn depend on requirements gathering. Testing and deployment are dependent on the coding phase. The chart helps ensure that resources focus on completing requirements and design first, allowing development to begin on schedule.

11.4.4 Detailed Example Using All Three Tools in a Software Project

Scenario – Developing a New E-Commerce Website

11.4.4.1 Flow Diagram

- A flow diagram could represent the user's journey, starting with "User Login," leading to "Product Browsing," followed by "Add to Cart," "Checkout," "Payment Processing," and ending at "Order Confirmation."
- This visual flow provides clarity on each user interaction point, guiding developers on which features need to be implemented sequentially.

11.4.4.2 Algorithmic Analysis

- **Search Algorithm** – For product browsing, algorithmic analysis might compare search algorithms to determine the best way to handle keyword-based product searches efficiently.
- **Payment Processing** – Evaluating cryptographic algorithms for data encryption to ensure security while optimizing for performance.
- By analysing and selecting efficient algorithms, the project ensures a balance between security and user experience in real-time operations.

11.4.4.3 Dependency Chart

- A dependency chart could show that "User Authentication" is a prerequisite for "User Login," which in turn must be complete before "Order History" can be accessed.
- Similarly, "Product Database Setup" is needed before implementing "Product Browsing," while "Payment Gateway Integration" depends on both "Checkout" and "User Authentication" modules.
- This helps organize development, assigning parallel tasks to different teams where possible and ensuring that critical dependencies are completed first to avoid delays.

11.4.5 Summary

- **Flow Diagrams** simplify understanding of processes, making them valuable for communication and troubleshooting.
- **Algorithmic Analysis** helps optimize performance and ensures efficient use of resources, which is essential for handling large data or high-demand operations.
- **Dependency Charts** are crucial for project planning, managing resources, and understanding task sequences, reducing the risk of delays or bottlenecks.

Together, these tools provide a structured approach to managing, analysing, and optimizing processes and projects.

CHAPTER TWELVE

12.1 STUDYING COMPLEX SYSTEMS

Complex systems are systems composed of many interconnected parts that exhibit emergent behavior which are behaviours that arise from the interaction of components rather than from the properties of the individual parts. These systems are prevalent in many fields, from natural sciences to social sciences, engineering, economics, and technology. Understanding complex systems involves studying how individual components interact, how these interactions lead to system-level behavior, and how the system responds to internal and external changes.

12.1.1 Key Concepts in Studying Complex Systems

1 Emergence

- ***Definition*** – Emergence is the phenomenon where larger patterns, properties, or behaviours arise from simple interactions of smaller components. These emergent behaviours are not always predictable from the individual parts.
- ***Example*** – In an ant colony, individual ants follow simple rules, but collectively they display complex behaviours like building nests, foraging, and organizing work, which can't be easily predicted by observing a single ant.

2 Nonlinearity

- ***Definition*** – Nonlinear systems are those where a change in one component does not lead to a proportional change in the system's behavior. Small changes can have large impacts, or large changes might result in minimal effects.
- ***Example*** – In financial markets, a small event, such as a rumour, can lead to a massive market crash or surge due to the nonlinear nature of investor behavior and interconnected financial instruments.

3 Feedback Loops

- ***Definition**:* Feedback loops are mechanisms where the output of a system influences its own input, either amplifying (positive feedback) or dampening (negative feedback) the behavior.

- ***Example*** – In ecosystems, predator-prey dynamics involve feedback loops where an increase in prey leads to more predators, which eventually reduces the prey population, leading to fewer predators.

4 Adaptation and Self-Organization

- ***Definition*** – Many complex systems adapt to changes in their environment and self-organize without central control. Components interact and adjust dynamically, leading to an organized system state.
- ***Example*** – In social networks, individuals form communities or clusters based on interactions, interests, or connections, leading to self-organization of the network without a single authority dictating the structure.

5 Interconnectedness and Dependencies

- ***Definition*** – Complex systems are highly interconnected, meaning that changes in one part can affect multiple other parts. This interconnectedness makes predicting outcomes challenging.
- ***Example*** – In a power grid, a failure in one power plant can lead to cascading failures across the grid, demonstrating the interconnected nature of the system.

6 Dynamical Systems

- ***Definition*** – Complex systems are often studied using dynamical systems theory, which focuses on how systems evolve over time. This includes understanding how systems change states, what influences stability, and what causes transitions between different states.
- ***Example*** – Weather systems are dynamical, where small atmospheric changes can lead to entirely different weather patterns (the "butterfly effect").

12.1.2 Methods for Studying Complex Systems

1 Modeling and Simulation

- ***Purpose*** – Create simplified models of complex systems to study their behavior. Simulations help test hypotheses, predict system behavior, and analyse possible scenarios.
- ***Tools*** – Agent-based models, Monte Carlo simulations, differential equations, and computer simulations.
- ***Example*** – Climate models simulate the Earth's climate using various variables, like temperature, humidity, and greenhouse gas levels, to predict future climate scenarios.

2 **Network Analysis**

- ***Purpose*** – Analyse the relationships and interactions between the components of a system. It helps in understanding connectivity, identifying influential nodes, and exploring system robustness.
- ***Tools*** – Graph theory, social network analysis, centrality measures, and clustering algorithms.
- ***Example*** – Studying the spread of diseases through social networks to identify critical nodes (individuals) for targeted vaccinations or interventions.

3 **Statistical and Computational Analysis**

- ***Purpose*** – Use data-driven methods to analyse complex systems, uncover patterns, identify correlations, and find anomalies.
- ***Tools*** – Machine learning, statistical analysis, Bayesian networks, regression models, and principal component analysis.
- ***Example*** – Using machine learning to analyse traffic flow data to predict congestion points and optimize traffic light patterns.

4 **Systems Dynamics**

- ***Purpose*** – Focuses on feedback loops, stock-and-flow structures, and time delays in systems to understand the behavior over time. It often includes creating causal loop diagrams.
- ***Tools*** – Systems dynamics software like Vensim or Stella, and mathematical modeling.
- ***Example*** – Analysing the population dynamics of an ecosystem by modeling predator-prey interactions over time.

5 **Chaos Theory and Sensitivity Analysis**

- ***Purpose*** – Study systems that appear random but have underlying patterns and are highly sensitive to initial conditions. This approach is used to understand unpredictable yet deterministic systems.
- ***Tools*** – Fractals, strange attractors, Lyapunov exponents, and bifurcation analysis.
- ***Example*** – Studying chaotic behavior in electrical systems or financial markets where small changes in parameters can lead to vastly different outcomes.

6 **Multiscale Analysis**

- ***Purpose*** – Study systems at different levels or scales, from micro (individual components) to macro (overall system). It's useful for understanding how small-scale interactions lead to large-scale patterns.
- ***Tools*** – Multi-resolution analysis, scale-space representation, and hierarchical models.
- ***Example*** – In biology, studying gene interactions (micro) and how they influence the overall behavior of an organism (macro).

12.1.3 Challenges in Studying Complex Systems

1 **Unpredictability** – Complex systems often exhibit unpredictable behavior due to nonlinearity and feedback loops.

- ***Solution*** – Use probabilistic models and simulations to explore multiple scenarios instead of deterministic outcomes.

2 **High Dimensionality** – Complex systems can involve a large number of variables and interactions, making analysis challenging.

- ***Solution*** – Dimensionality reduction techniques like principal component analysis (PCA) or clustering to simplify analysis.

3 **Emergent Behavior** – Emergent phenomena make it difficult to predict system behavior by examining individual parts.

- ***Solution*** – Use agent-based models and simulations that focus on interactions rather than individual behavior.

4 **Interdependencies** – Systems may have hidden dependencies where components affect each other in unexpected ways.

- ***Solution***: Employ network analysis and system dynamics to identify critical dependencies and potential weak points.

5 **Adaptation and Evolution** – Complex systems can evolve over time, making static models obsolete.

- ***Solution*** – Use adaptive models and real-time data to continuously update and refine the understanding of the system.

12.1.4 Real-World Examples of Complex Systems

1 **Traffic Systems** – Urban traffic flow is a complex system involving vehicles, pedestrians, infrastructure, weather, and human behavior. Understanding and predicting traffic patterns requires analysing multiple interconnected variables.

- ***Solution*** – Traffic simulation models like SUMO (Simulation of Urban Mobility) help city planners optimize road usage and reduce congestion.

2 **Healthcare Systems** – Healthcare involves a complex web of patients, healthcare providers, insurance systems, pharmaceuticals, and government regulations. Managing healthcare effectively requires understanding this interconnected ecosystem.

- ***Solution*** – Network analysis of patient data and simulations of disease spread help in resource allocation and policy-making.

3 **Ecosystems** – Ecosystems are highly complex systems with diverse species interacting with each other and their environment. Changes in one species or environmental factor can have cascading effects.

- ***Solution*** – Ecological modeling using tools like Ecopath or Stella helps in conservation planning and understanding the impact of human activities on biodiversity.

12.1.5 Summary

In summary, studying complex systems requires a multi-disciplinary approach that includes modeling, simulation, network analysis, and statistical tools. It demands a mindset that embraces uncertainty, nonlinearity, and emergent phenomena. With advances in computational power, data analysis, and algorithmic techniques, our ability to analyse and understand complex systems continues to improve, leading to better decision-making in fields ranging from engineering and economics to biology and urban.

12.1.6 Multiple Problem Analysis

Multiple Problem Analysis involves systematically identifying, analysing, and prioritizing several problems within a system, project, or environment. It helps understand how different issues may interact, overlap, or impact each other. This process is critical in complex scenarios where problems are interconnected, and solutions must be integrated to be effective. Importance for multiple problem analysis is enumerated as follows:

- **Interconnected Issues** – Many problems are not isolated; they are linked with other issues. Addressing one without considering its relationship to others can lead to unintended consequences.

- **Prioritization** – Understanding the severity and impact of each problem helps in setting priorities, allocating resources, and developing a strategic approach.
- **Comprehensive Solutions** – A holistic view of multiple problems allows for the creation of more robust and integrated solutions.

12.1.7 Steps in Multiple Problem Analysis

1 Problem Identification

- List and identify all existing problems within the system or project.
- Use tools like brainstorming, surveys, expert opinions, or field observations to gather information.
- ***Example*** – In a factory setting, problems might include equipment downtime, quality defects, high energy consumption, and safety incidents.

2 Problem Categorization

- Group similar problems to see if they share common causes or characteristics.
- Categorize problems into different types, such as technical, managerial, operational, or environmental.
- ***Example*** – In a software project, problems could be categorized into bugs, performance issues, user interface problems, and integration challenges.

3 Prioritization

- Rank the problems based on their severity, impact, cost, or urgency.
- Use tools like **Pareto Analysis** (80/20 rule) or **Weighted Scoring** to prioritize.
- ***Example*** – In a construction project, prioritizing issues like structural integrity concerns over minor aesthetic problems due to safety risks.

4 Cause Analysis

- Determine the root causes of each problem using techniques like **Root Cause Analysis (RCA), Fishbone Diagrams (Ishikawa)**, or **5 Whys**.
- Identify if multiple problems share a common root cause, suggesting that solving one issue may address others.
- ***Example*** – In a supply chain scenario, analysing whether delays, high costs, and stockouts share common causes like supplier inefficiencies or transportation bottlenecks.

5 **Impact Analysis**

- Assess the impact of each problem on the overall system, including direct and indirect consequences.
- Consider using **Failure Mode and Effects Analysis (FMEA)** to evaluate the severity, occurrence, and detection of each issue.
- ***Example*** – In a hospital, evaluating the impact of multiple problems like equipment failures, staff shortages, and patient complaints on patient care quality and safety.

6 **Solution Exploration**

- Explore potential solutions for each problem, considering both short-term fixes and long-term strategies.
- Develop multiple options, evaluate their feasibility, and estimate costs.
- ***Example*** – In a data center, considering solutions like upgrading hardware, optimizing software configurations, and training staff to handle issues related to server downtime and poor performance.

7 **Integration of Solutions**

- Look for integrated solutions that address multiple problems simultaneously.
- Develop an implementation plan that sequences the solutions logically to maximize impact.
- ***Example*** – In urban planning, implementing a public transport system may simultaneously reduce traffic congestion, lower pollution levels, and improve accessibility.

8 **Monitoring and Evaluation**:

- Track the progress of the solutions using Key Performance Indicators (KPIs) or metrics.
- Re-assess if new problems emerge or if the initial issues persist, requiring adjustments to the strategy.
- ***Example*** – In a manufacturing plant, monitoring the impact of new maintenance protocols on equipment reliability, quality, and production efficiency.

12.1.8 Examples of Multiple Problem Analysis in Different Industries

12.1.8.1 Manufacturing Industry

- **Scenario** – A manufacturing plant faces multiple problems — increased production downtime, low product quality, safety incidents, and high operational costs.

Analysis

- **Categorize** – Categorize problems as **Operational** (downtime, costs), **Quality** (product defects), and **Safety** (incidents).
- **Prioritize** – Address safety incidents first due to their impact on worker safety, then focus on quality issues, followed by downtime and cost problems.
- **Root Causes** – Use Fishbone Diagrams to identify root causes like poor machine maintenance, lack of training, and substandard raw materials.
- **Solution** – Implement better training programs, upgrade maintenance schedules, improve supplier quality control, and introduce energy-saving practices to reduce costs.
- **Integrated Solution** – A comprehensive quality management system could address safety, quality, and operational efficiency simultaneously.

12.1.8.2 Healthcare Industry

- **Scenario** – A hospital deals with high patient wait times, staff shortages, equipment failures, and patient dissatisfaction.

Analysis

- ***Categorize*** – Issues related to **Operational Efficiency** (wait times), **Staffing** (shortages), **Technology** (equipment failures), and **Patient Care** (satisfaction).
- ***Prioritize*** – Focus on equipment reliability to ensure critical care, then address staff shortages, as they have a direct impact on patient satisfaction.
- ***Root Causes*** – Identify root causes using the 5 Whys, revealing inadequate hiring processes, insufficient equipment maintenance, and scheduling inefficiencies.
- ***Solution*** – Hire additional staff, introduce preventative maintenance for equipment, and optimize scheduling to reduce wait times.
- **Integrated Solution** – Implement a comprehensive hospital management system to track maintenance, manage staffing, and schedule appointments, improving efficiency across the board.

12.1.8.3 IT and Software Development

- **Scenario** – A software development team faces multiple problems — frequent bugs, missed deadlines, high client dissatisfaction, and poor performance.

Analysis

- ***Categorize*** – Issues can be broken down into **Technical** (bugs, performance), **Process** (missed deadlines), and **Customer Relations** (client dissatisfaction).
- ***Prioritize*** – Fix bugs and performance issues first to stabilize the software, and then focus on meeting project deadlines.
- ***Root Causes*** – Use Root Cause Analysis to identify underlying problems like lack of documentation, insufficient testing, and poor communication.
- ***Solution*** – Introduce test-driven development, improve project management practices, and establish clear client communication channels.
- **Integrated Solution** – Agile project management tools and better version control systems can help in coordinating development, improving timelines, and reducing bugs.

12.1.8.4 Construction Industry

- **Scenario** – A construction project faces challenges such as budget overruns, delays, safety violations, and stakeholder disputes.

Analysis

- ***Categorize*** – Issues can be grouped into **Financial** (budget), **Scheduling** (delays), **Safety** (violations), and **Stakeholder Management** (disputes).
- ***Prioritize*** – Focus on safety violations first to comply with regulations, then address scheduling issues to control budget overruns.
- ***Root Causes*** – Using a Fishbone Diagram, root causes could include poor planning, lack of communication, and insufficient safety training.
- ***Solution*** – Strengthen safety protocols, improve project planning, enhance communication among stakeholders, and conduct regular budget reviews.
- **Integrated Solution** – Implementing project management software and a safety management system can streamline processes and improve compliance.

12.1.9 Tools for Multiple Problem Analysis

1. **SWOT Analysis** – Identifies Strengths, Weaknesses, Opportunities, and Threats related to multiple issues.
2. **Pareto Analysis** – Focuses on the vital few problems that cause the majority of the impact.
3. **Fishbone Diagram (Ishikawa)** – Helps in identifying the root causes of problems in a structured way.
4. **5 Whys** – A simple technique to drill down to the root cause of each problem.
5. **Decision Matrix** – Helps evaluate and prioritize potential solutions based on predefined criteria.
6. **Failure Mode and Effects Analysis (FMEA)**: Assesses the impact, severity, and detectability of multiple problems to prioritize actions.

12.1.10 Summary

In summary, Multiple Problem Analysis is essential for understanding and addressing complex, interconnected challenges in any system or industry. By systematically identifying, analysing, and integrating solutions, it is possible to achieve outcomes that are more effective. This approach promotes comprehensive problem solving, minimizes unintended consequences, and helps optimize resources for long-term success.

12.2 TROUBLESHOOTING AND DIAGNOSIS OF FAULTS IN INDUSTRIAL EQUIPMENT AND MACHINERY

Troubleshooting and diagnosing faults in industrial equipment and machinery are essential to maintaining operational efficiency, minimizing downtime, and ensuring safety. This process involves a systematic and logical approach to identifying problems, understanding their root causes, and implementing effective solutions. Given the complexity of modern industrial systems, accurate diagnosis requires a combination of technical skills, specialized tools, and a good understanding of how the machinery operates.

12.2.1 General Steps for Troubleshooting and Diagnosis

1 Information Gathering

- ***Purpose*** – Understand the nature and symptoms of the problem. Gather as much data as possible to define the issue.

Actions

- Collect operational data from the machinery.

- Interview operators or technicians who noticed the fault.
- Review logs, error messages, or diagnostic reports.

- ***Example*** – In a CNC machine, if the precision of cuts is off, gather information on recent machine parameters, tool settings, and maintenance logs.

2 Problem Definition

- **Purpose** – Clearly define what the fault is and what part of the machinery is affected.

Actions

- Identify the symptoms, their location, and any observable trends.
- Document abnormal behavior, sounds, vibrations, temperatures, or performance changes.

- ***Example*** – If a conveyor belt is malfunctioning, determine if the fault is related to the motor, belt tension, misalignment, or other mechanical components.

3 Initial Inspection

- **Purpose** – Perform a quick visual and operational check to identify obvious causes.

Actions

- Conduct a visual inspection of the equipment to look for visible damage, leaks, or loose components.
- Test the machine in safe operating conditions to observe any anomalies.

- ***Example*** – In a hydraulic press, check for hydraulic fluid leaks, pressure gauge readings, and unusual noises during operation.

4 Perform Diagnostics

- **Purpose** – Use diagnostic tools and techniques to collect more precise information about the fault.

Actions

- Use sensors, meters, or diagnostic software to collect data.
- Run built-in diagnostic tests, if available, for specific machines.
- Use thermal imaging, vibration analysis, or ultrasonic testing to identify hidden faults.

- ***Example*** – In a gas turbine, use vibration analysis to check for imbalances in the rotating components or perform an oil analysis to detect signs of wear.

5 Hypothesis Formation

- ***Purpose:*** Formulate possible causes based on the gathered data and diagnostics.

Actions

- Use historical data, technical manuals, and expert knowledge to develop hypotheses.
- Prioritize potential causes from most likely to least likely.

- ***Example*** – If an industrial boiler is not heating efficiently, consider potential causes like faulty burners, clogged filters, poor fuel quality, or issues with the heat exchanger.

6 Testing and Analysis

- **Purpose** – Test each hypothesis to confirm or rule out possible causes.

Actions

- Make adjustments or temporary fixes to test each potential cause.
- Use a process of elimination to narrow down the root cause.

- ***Example*** – In a robotic welding system, if the weld quality is poor, check if the problem is due to improper calibration, worn-out electrodes, or an issue with the power supply.

7 Root Cause Identification

- **Purpose** – Identify the primary cause of the fault.

Actions

- Use techniques like **Root Cause Analysis (RCA)** to determine the underlying issue.
- Identify if there are multiple contributing factors.

- ***Example*** – In an air compressor, if the output pressure is low, RCA might reveal a clogged air filter, worn piston rings, or leaks in the system.

8 Implementation of Solution

- **Purpose** – Apply the appropriate corrective measures to fix the problem.

Actions

- Replace faulty parts, recalibrate machinery, or make adjustments based on the diagnosis.
- Follow manufacturer guidelines for repair and maintenance.

- ***Example*** – In an electrical motor showing signs of overheating, replace worn bearings, lubricate moving parts, or repair damaged windings.

9 Verification and Testing

- **Purpose** – Ensure that the problem is resolved and the machinery operates correctly.

Actions

- Test the equipment under normal operating conditions.
- Monitor for recurrence of symptoms or new issues.

- ***Example*** – After replacing a valve in a fluid pump, check the pressure and flow rates to ensure they meet specifications.

10 Documentation and Reporting

- **Purpose** – Keep records of the problem, diagnosis process, and the solutions applied.

Actions

- Document findings, actions taken, and lessons learned for future reference.
- Update maintenance logs and create reports for stakeholders.

- ***Example*** – In an industrial lathe repair, document the part replaced, diagnostic steps, solution applied, and any recommendations for future maintenance.

12.2.2 Examples of Troubleshooting and Diagnosis in Specific Industries

1 Manufacturing - CNC Machines

- **Problem** – CNC machine exhibits poor cutting precision.

Diagnosis

- Check if the tool is properly aligned and secured.
- Inspect for tool wear or damage.
- Run diagnostics to check the axis calibration and control system integrity.
- **Solution** – Realign the tool, replace worn-out parts, and recalibrate the machine's axis.
- **Outcome** – Improved precision, reduced scrap, and consistent quality.

2 Power Generation - Gas Turbines

- **Problem** – Unexpected drop in efficiency and increased vibration in a gas turbine.

Diagnosis

- Perform a thermal imaging scan to detect abnormal temperature variations.
- Use vibration analysis to identify imbalances or misalignments.

- Check for any signs of wear in the bearings or rotor blades.
- **Solution** – Rebalance the rotating parts, replace damaged bearings, and inspect for any fouling in the combustion chamber.
- **Outcome** – Restored turbine efficiency and reduced vibration levels.

3 **Oil and Gas - Pumps and Compressors**

- **Problem** – A pump shows reduced output pressure and flow rate.

Diagnosis

- Inspect seals, gaskets, and pipes for leaks.
- Measure pressure at different points using pressure gauges.
- Check the condition of the impeller and bearings.
- **Solution** – Replace worn seals, clean or replace the impeller, and tighten loose connections.
- **Outcome** – Normalized flow rates and improved system efficiency.

4 **Food Processing - Conveyor Systems**

- **Problem** – Conveyor belt stalls intermittently.

Diagnosis

- Check for proper tension and alignment of the belt.
- Inspect the motor, gearbox, and rollers for wear or damage.
- Examine control system sensors for faults.
- **Solution** – Adjust belt tension, replace damaged motor parts, and recalibrate sensors.
- **Outcome** – Continuous and smooth conveyor operation.

12.2.3 Common Tools for Troubleshooting and Diagnosis

1 **Multimeter** – For measuring voltage, current, and resistance in electrical systems.

- **Example** – Diagnosing electrical faults in motors or control circuits.

2 **Vibration Analyser** – To detect misalignment, imbalance, or wear in rotating equipment.

- **Example** – Used in diagnosing bearing faults in industrial fans.

3 **Thermal Imaging Camera** – Detects temperature variations and hotspots.

- **Example** – Used to find overheating in electrical panels or mechanical components.

4 **Ultrasonic Detector** – Identifies leaks in pipes, air systems, or steam lines.

- **Example** – Detecting leaks in compressed air systems.

5 **PLC Diagnostic Software** – To check for faults in Programmable Logic Controllers (PLCs) and control systems.

- **Example** – Troubleshooting control system failures in automated machinery.

6 **Pressure Gauges** – To measure pressure levels in pneumatic and hydraulic systems.

- **Example** – Detecting pressure drops in hydraulic presses.

7 **Diagnostic Software** – For accessing error logs, system statuses, and detailed operational data.

- **Example** – Used in CNC machines to diagnose software-related issues.

12.2.4 Challenges in Troubleshooting and Diagnosis

1 **Complex Systems** – Modern industrial equipment is increasingly complex, making it difficult to pinpoint faults without specialized knowledge or tools.

- **Solution** – Invest in advanced diagnostic tools and provide regular training for technicians.

2 **Intermittent Issues** – Some problems only appear intermittently, making diagnosis challenging.

- **Solution** – Use continuous monitoring systems to capture data during faults for analysis.

3 **Lack of Documentation** – Inadequate or missing maintenance records can complicate fault analysis.

- **Solution** – Keep detailed and updated maintenance logs to aid future troubleshooting.

4 **Multiple Faults** – Simultaneous issues may overlap, masking the real root cause.

- **Solution** – Address the most likely problems first, and use a process of elimination for accurate diagnosis.

12.2.5 Summary

In summary, effective troubleshooting and diagnosis of industrial equipment faults are crucial for minimizing downtime, reducing maintenance costs, and enhancing productivity. By following a systematic approach, using proper diagnostic tools, and understanding the principles of fault analysis, technicians can maintain reliable and efficient operations.

Documentation, training, and technology play vital roles in facilitating accurate and timely fault resolution, ensuring that the industrial systems remain safe and productive.

CHAPTER THIRTEEN

13.0 MATERIALS HANDLING EQUIPMENT

Materials handling equipment plays a crucial role during repairs in industrial settings, aiding in the safe, efficient, and effective movement of tools, parts, and heavy machinery. When equipment breaks down or needs maintenance, the right materials handling tools are essential to facilitate repairs, minimize downtime, and ensure safety.

13.1 Materials Handling Equipment Commonly Used During Repairs

13.1.1 Lifting and Hoisting Equipment

- **Purpose** – To lift, lower, or reposition heavy machinery or components during repairs. These tools are critical when disassembling or accessing parts that are not easily reachable.

Examples

- **Cranes** – Overhead cranes or mobile cranes are used to lift and move heavy equipment like engines, motors, or other machinery parts that require repair.
  - ***Example*** – An overhead crane may be used to lift a malfunctioning industrial pump off its mount to facilitate repairs.
- **Hoists**: Chain hoists, electric hoists, or lever hoists allow for precise lifting of smaller components during disassembly.
  - ***Example*** – A chain hoist can be used to remove a heavy gear assembly from an engine for maintenance.
- **Jacks** – Hydraulic or mechanical jacks are used to raise machinery to access components beneath.
  - ***Example*** – A hydraulic jack can lift a forklift to perform repairs on the undercarriage or tires.

13.1.2 Conveyors and Rollers

- **Purpose** – To transport heavy parts, tools, or equipment from one area to another within a facility during repairs.

Examples

- o **Portable Conveyors** – Belt or roller conveyors that can be moved to the repair area to carry tools, replacement parts, or debris.
- ▪ ***Example*** – A portable roller conveyor might be used to transport heavy replacement parts from storage to the repair site.
- o **Skates and Rollers** – Rollers are placed under machinery or equipment to move them across a flat surface easily.
- ▪ ***Example*** – Machinery skates can be positioned under a heavy machine to relocate it to a repair bay without using a crane.

13.1.3 Transport and Carrying Equipment

- **Purpose** – To move components, tools, and supplies to the repair site efficiently.

Examples

- o **Forklifts** – Forklifts are essential for transporting bulky items, spare parts, or even broken equipment to the maintenance area.
- ▪ ***Example*** – A forklift can transport a heavy gearbox or a new component to replace a damaged one on the factory floor.
- o **Pallet Jacks** – Used to move pallets loaded with spare parts or tools to and from the repair area.
- ▪ ***Example*** – A pallet jack can be used to bring a pallet of tools, lubricants, or spare parts closer to the machine being serviced.
- o **Tool Carts** – Rolling carts with multiple compartments for organizing and transporting tools and repair supplies.
- ▪ ***Example*** – A mechanic may use a tool cart to carry all necessary wrenches, screwdrivers, bolts, and diagnostics tools to the repair site.

13.1.4 Positioning Equipment

- **Purpose** – To hold, position, or manipulate components during repair work to improve accessibility and accuracy.

Examples

- o **Lift Tables** – Adjustable height platforms to raise or lower equipment to a comfortable working level for repairs.

- *Example* – A lift table may be used to elevate a transmission component to an ergonomic height for maintenance.

- **Workbenches with Vises** – Heavy-duty workbenches equipped with vises to hold components securely during repairs.

- *Example* – A workbench can secure a small engine part in a vise to facilitate repairs, like tightening or welding.

- **Turntables** – Rotating platforms that allow workers to access different sides of a heavy item without repositioning manually.

- *Example* – A technician might use a turntable to rotate a large pump for a detailed inspection.

13.1.5 Storage and Retrieval Equipment

- **Purpose** – To organize, retrieve, and deliver components, tools, and spare parts needed during repairs.

Examples

- **Storage Bins and Cabinets** – Used to store smaller components, fasteners, or specialized tools needed for repairs.

- *Example* – A maintenance technician may access a storage bin to retrieve a replacement bearing or set of screws.

- **Shelving Systems** – Open or closed shelving units that hold larger components or tools.

- *Example* – A mechanic retrieves a specialized wrench from a tool cabinet to work on an industrial machine.

- **Retractable Tool Balancers** – Tools are hung from retractable balancers, allowing easy access during repairs without clutter.

- *Example* – A welder might have a torch suspended from a tool balancer, keeping it within reach during repairs.

13.1.6 Handling Equipment for Hazardous Materials

- **Purpose** – To safely manage and handle potentially dangerous materials or equipment during repairs.

Examples

- **Drum Lifters and Rotators** – Used to move and position drums containing chemicals, lubricants, or other materials.

- *Example* – A drum lifter might be used to rotate and empty a drum of hazardous chemicals safely during machinery servicing.

- **Spill Containment Pallets** – Pallets designed to contain spills, often used when handling or moving equipment with potential fluid leaks.

- *Example* – A containment pallet is used to hold a leaky hydraulic pump during disassembly for repairs.

13.1.7 Portable Scaffolding and Ladders

- **Purpose** – To provide access to high or hard-to-reach areas during machinery repair or maintenance.

Examples

- **Portable Scaffolding** – Temporary structures that provide stable platforms for workers repairing overhead machinery or equipment.

- *Example* – Scaffolding may be erected to access high conveyor belts for repair or maintenance.

- **Mobile Ladders** – Wheeled ladders that can be easily moved to different areas for repairs.

- *Example* – A mobile ladder may be used to reach a high component of an industrial machine that needs inspection or repair.

13.1.8 Protective Handling Equipment

- **Purpose** – To ensure safe handling of equipment or parts that may have sharp edges, high temperatures, or dangerous chemicals.

Examples

- **Heat-Resistant Gloves** – Used for handling hot machinery parts during repairs.

- *Example* – A mechanic may wear heat-resistant gloves when handling a machine part that was recently shut down.

- **Protective Mats** – Placed under equipment to catch debris or fluids during disassembly and repair.

- *Example* – A rubber mat might be used to protect the floor while dismantling a leaking hydraulic system.

Figure 13.1: Rubber Mat

13.1.9 Benefits of Using Materials Handling Equipment During Repairs

1 **Efficiency** – Speeds up repair processes by enabling the easy movement and access of parts.

2 **Safety** – Reduces the risk of injury by allowing machinery and heavy items to be handled safely.

3 **Cost-Effectiveness** – Minimizes downtime and labor costs by making repairs more efficient.

4 **Improved Accuracy** – Provides stable and ergonomic platforms for precise maintenance and repairs.

In summary, the appropriate use of materials handling equipment during repairs not only increases safety and efficiency but also extends the lifespan of industrial machinery by ensuring that maintenance tasks are carried out with precision and care.

13.2 SAFETY RULES FOR MATERIALS HANDLING JOBS

Safety is crucial in materials handling jobs to prevent injuries, damage to equipment, and ensure a smooth workflow. Proper safety practices reduce the risk of accidents when lifting, transporting, storing, or handling materials.

13.2.1 Key Safety Rules for Materials Handling Jobs With Examples

1 Proper Training and Certification

- **Rule** – Workers must receive training on how to safely operate materials handling equipment, understand the hazards involved, and follow safe handling procedures.

Examples

- o **Forklift Certification** – Operators must complete certified training programs to operate forklifts, focusing on load limits, maneuvering, and parking safely.

- **Crane Operation** – Crane operators should be trained to read load charts, handle suspended loads, and communicate using standard hand signals.

2 Inspect Equipment Before Use

- **Rule** – Always perform a pre-use inspection of materials handling equipment to ensure it is in good condition. Any equipment that appears damaged or defective should not be used.

Examples

- **Forklift Inspection** – Inspect forks for cracks, tires for wear, the hydraulic system for leaks, and brakes for proper function before starting.
- **Conveyor Belt Inspection** – Check belts for tension and damage, ensure safety guards are in place, and examine moving parts for proper lubrication.

3 Wear Appropriate Personal Protective Equipment (PPE)

- **Rule** – Workers must wear the correct PPE to protect against specific hazards associated with materials handling tasks.

Examples

- ***Hard Hats:*** Worn to protect against falling objects when stacking materials on high shelves or operating cranes.
- ***Steel-Toe Boots*** – Essential for protecting feet from heavy falling objects during loading or unloading.
- ***Gloves*** – Use heavy-duty gloves when handling rough, sharp, or abrasive materials like metal sheets or broken pallets.

4 Follow Load Limits and Stability Guidelines

- **Rule** – Always adhere to equipment load limits to prevent tipping, falling loads, or mechanical failure.

Examples

- **Forklift Load Capacity** – Never exceed the forklift's weight capacity, and keep the load centered and low to maintain stability.
- **Manual Lifting** – When manually lifting objects, never lift more than the recommended weight (typically 50 lbs/22 kg for a single person) and use proper lifting techniques.

5 **Use Proper Lifting Techniques**

- **Rule** – Workers should use ergonomic and safe lifting techniques to avoid injuries, especially back injuries.

Examples

- ***Bend Knees, Not Back*** – When lifting a heavy object, squat down, keep the back straight, grip firmly, and lift using the legs, not the back.
- ***Team Lifting*** – For heavy or bulky items, ask for assistance or use team lifting to share the load equally.

6 **Secure and Balance Loads Properly**

- **Rule** – Ensure loads are securely tied down, balanced, and stable before lifting or moving them.

Examples

- ***Pallet Stacking*** – Stack items evenly on pallets, ensuring they are not leaning, and secure them with shrink wrap or straps.
- ***Forklift Handling*** – Center the load on the forks, tilt the mast slightly back, and move slowly to prevent tipping or sliding.

7 **Clear Pathways and Maintain Good Housekeeping**

- **Rule** – Always keep pathways clear of obstructions and ensure the workspace is free from debris, spills, or tripping hazards.

Examples

- ***Conveyor Areas*** – Keep conveyor belts free from loose items or debris that could jam or damage the system.
- ***Loading Docks*** – Clear any clutter or obstacles that could block loading/unloading areas, ensuring smooth and safe operations.

8 **Use Appropriate Equipment for the Task**

- **Rule** – Use the correct type of materials handling equipment for the specific job to ensure safety and efficiency.

Examples

- ***Use Dollies for Heavy Loads*** – Instead of manually carrying heavy objects, use a dolly or hand truck to minimize the risk of injury.

- *Overhead Crane for Heavy Machinery* – Use cranes for moving heavy machinery instead of using a forklift to avoid strain and overloading.

9 Communicate and Use Hand Signals

- **Rule** – Communication between workers, especially in noisy environments, is vital. Use standardized hand signals, radios, or other methods to ensure safe handling.

Examples

- *Crane Operation* – A spotter should use standard hand signals to guide the crane operator when moving large loads.
- *Team Lifting* – Clearly communicate the timing and direction when moving heavy or awkward items as a team.

10 Lockout/Tagout (LOTO) for Repairs and Maintenance

- **Rule** – Ensure equipment is completely powered off and tagged out before performing maintenance or repairs to prevent accidental startup.

Examples

- *Conveyor Belt Maintenance* – Use LOTO procedures to disconnect power to a conveyor system before replacing a belt or motor.
- *Forklift Repairs* – Tag out a forklift under maintenance to ensure no one accidentally starts or moves it.

11 Avoid Overreaching and Maintain Balance

- **Rule** – Avoid reaching too far when handling materials or operating equipment, as it can lead to loss of balance or strains.

Examples

- *On Ladders* – Avoid leaning too far to the side when on a ladder; reposition the ladder instead.
- *Manual Lifting* – Avoid twisting the body when lifting or lowering loads; instead, move the feet to maintain a stable stance.

12 Follow Proper Loading and Unloading Procedures

- **Rule** – Use safe loading and unloading practices to prevent accidents and injuries.

Examples

- ***Loading a Truck*** – Distribute weight evenly and secure cargo with straps or tie-downs to prevent shifting during transport.
- ***Unloading Pallets*** – Carefully remove straps and secure items before moving or lifting to avoid falling materials.

13 Observe Safe Speeds and Manoeuvring

- **Rule** – Operators should follow speed limits and maintain safe driving practices when operating materials handling equipment like forklifts or AGVs.

Examples

- ***Forklift Operation:*** Drive slowly in crowded areas, use the horn at intersections, and avoid sharp turns to prevent tipping.
- ***Pallet Jack Use*** – Move slowly and steadily when transporting heavy loads to maintain balance and avoid losing control.

14 Report and Address Hazards Immediately

- **Rule** – Encourage workers to report any hazards, unsafe conditions, or damaged equipment immediately to prevent accidents.
- **Examples**
- ***Damaged Racking*** – If warehouse racking appears damaged or unstable, it should be reported and repaired before use.
- ***Leaking Hydraulic Fluid*** – If a forklift has a hydraulic leak, it should be reported and removed from service until repaired.

13.2.2 Key Takeaways for Safety in Materials Handling

- Safety in materials handling jobs involves using the right equipment, following proper procedures, and paying attention to details.
- Emphasis on training, inspections, PPE, communication, and correct equipment use minimizes risks.
- Adhering to safety guidelines not only protects workers but also reduces costs associated with accidents, equipment damage, and downtime.

When these safety rules and practices are applied, workplaces can create a safer and more efficient environment for handling materials, ensuring that workers and equipment remain productive and protected.

CHAPTER FOURTEEN

14.0 ENGINEERING SERVICES

Engineering Services encompass a wide range of professional services provided by engineers to design, develop, maintain, and improve systems, structures, and equipment. These services span multiple industries and are essential for ensuring functionality, safety, efficiency, and innovation in various fields.

14.1.1 Common Engineering Services With Practical Examples

1 Design and Drafting Services

- **Purpose** – Engineers create detailed designs and plans for new products, buildings, systems, or infrastructure, using computer-aided design (CAD) software. These designs serve as blueprints for construction, manufacturing, or development.

Examples

- o **Architectural Design** – Structural engineers design bridges, residential buildings, and skyscrapers, ensuring they are safe, stable, and meet all regulatory standards.
- ▪ *Example* – A structural engineer designs a suspension bridge, calculating load capacities and materials to ensure safety.
- o **Mechanical Drafting** – Mechanical engineers draft detailed designs for manufacturing components like engines, turbines, or industrial machines.
- ▪ *Example* – A mechanical engineer creates CAD drawings for a custom gearbox used in heavy machinery.

2 Consulting Services

- **Purpose** – Engineering consultants provide expert advice and guidance to companies, helping them solve technical problems, optimize processes, comply with regulations, and develop effective strategies.

Examples

- o **Environmental Consulting** – Environmental engineers consult on projects to reduce waste, manage pollution, or develop sustainable practices.
- ▪ *Example* – An environmental engineer consults with a factory to implement air quality control systems and reduce emissions.

- **Energy Efficiency Consulting** – Engineers analyse energy consumption and recommend ways to reduce energy use in commercial or industrial settings.
  - ***Example*** – An energy engineer assesses an industrial plant's energy consumption and recommends upgrading to energy-efficient lighting and machinery.

3 Project Management

- **Purpose** – Engineers provide project management services to oversee the planning, execution, and completion of engineering projects, ensuring they stay on schedule, within budget, and meet quality standards.

Examples

- **Construction Project Management** – Civil engineers manage the construction of highways, bridges, or tunnels, coordinating resources, teams, and schedules.
  - ***Example*** – A civil engineer oversees the construction of a new highway, managing subcontractors, materials, and construction timelines.
- **Product Development Management** – Mechanical engineers manage the development of new products, from initial concept to final production, including prototyping and testing.
  - ***Example*** – A mechanical engineer leads the design and testing phases of a new medical device, ensuring it meets safety and functionality standards.

4 Inspection and Testing Services

- **Purpose**: Engineers conduct inspections and tests to ensure equipment, systems, and infrastructure meet safety, quality, and performance standards. These services are vital for identifying defects or issues before they become serious problems.

Examples

- **Structural Inspection** – Structural engineers inspect buildings, bridges, or dams for signs of wear, cracks, or structural integrity.
  - ***Example*** – A structural engineer conducts a safety inspection of a dam to identify potential cracks or weaknesses.
- **Material Testing** – Engineers test materials for strength, durability, and suitability for specific applications, using methods like tensile testing or fatigue analysis.
  - ***Example*** – A materials engineer tests the metal alloy used in aircraft construction to ensure it meets strength and weight requirements.

5 Maintenance and Repair Services

- **Purpose** – Engineers provide maintenance and repair services to keep systems, machines, or infrastructure running efficiently. This can include routine maintenance, emergency repairs, or system upgrades.

Examples

- **Industrial Maintenance** – Mechanical engineers maintain and repair manufacturing equipment, ensuring it operates at peak efficiency.
  - ***Example*** – An engineer performs preventive maintenance on an assembly line's robotic arms to avoid unexpected breakdowns.
- **Electrical Repairs** – Electrical engineers troubleshoot and repair electrical systems in buildings, factories, or power plants.
  - ***Example*** – An electrical engineer repairs a malfunctioning circuit breaker in a manufacturing facility to restore power.

6 Construction Services

- **Purpose** – Engineers oversee the actual construction of infrastructure, buildings, or industrial projects, ensuring that work follows design specifications, regulations, and quality standards.

Examples

- **Civil Construction** – Civil engineers manage the construction of roads, bridges, and utility systems, including excavation, foundation laying, and paving.
  - ***Example*** – A civil engineer supervises the construction of a multi-story parking garage, ensuring structural stability and adherence to local building codes.
- **Electrical Installation** – Electrical engineers handle the installation of electrical systems, including wiring, lighting, and electrical panels.
  - ***Example*** – An electrical engineer installs and configures the electrical grid in a new industrial park.

7 Research and Development (R&D)

- **Purpose**: Engineers are involved in R&D to innovate and develop new technologies, products, or processes that improve efficiency, reduce costs, or solve technical challenges.

Examples

- **Product Development** – Engineers work on designing new products, from automotive components to consumer electronics, by experimenting with new materials, techniques, and technologies.
- ***Example*** – An automotive engineer develops a prototype for a hybrid engine that increases fuel efficiency.
- **Process Optimization** – Engineers research ways to make manufacturing or industrial processes more efficient, sustainable, or cost-effective.
- *Example* – A chemical engineer works on optimizing the formula for an eco-friendly detergent that uses biodegradable ingredients.

8 Energy Management Services

- **Purpose** – Engineers offer services related to energy management, focusing on renewable energy, efficiency, and the reduction of carbon footprints.

Examples

- **Renewable Energy Solutions** – Engineers design solar, wind, or hydroelectric systems to provide renewable energy for commercial or residential use.
- ***Example*** – An energy engineer designs a solar power system for a rural community, calculating panel placement and energy output.
- **Energy Audits** – Engineers perform audits of existing facilities to identify areas where energy use can be reduced, recommending upgrades or modifications.
- ***Example*** – An energy engineer conducts an audit of a manufacturing facility and suggests installing energy-efficient motors to reduce electricity costs.

9 Automation and Control Systems

- **Purpose** – Engineers develop automation systems to improve manufacturing efficiency, reduce manual labor, and enhance precision in industrial processes.

Examples

- **Robotic Automation** – Engineers design and implement robotic systems for repetitive or hazardous tasks in manufacturing.
- ***Example*** – A robotic engineer installs and programs a robot to perform quality checks on an assembly line.

- **Control Systems** – Engineers create control systems for automated machinery, like programmable logic controllers (PLCs) for industrial processes.
  - *Example* – An automation engineer develops a PLC system to control the temperature and pressure of a chemical reactor.

10 Environmental Engineering Services

- **Purpose** – Engineers develop solutions to manage and reduce environmental impact, focusing on sustainability, waste management, and pollution control.

Examples

- **Water Treatment Systems** – Engineers design systems to treat wastewater, ensuring it meets environmental standards before release.
  - *Example* – An environmental engineer designs a water filtration system for a factory to reduce pollutants discharged into local rivers.
- **Air Pollution Control** – Engineers develop and install equipment to limit air pollution from industrial sources.
  - *Example* – An environmental engineer installs scrubbers in a power plant to capture harmful emissions.

14.1.2 Importance of Engineering Services

1. **Innovation**: Engineers drive technological advancements and improve existing systems, making them more efficient and cost-effective.
2. **Safety** – Engineering services ensure that structures, machinery, and systems are safe for public use, preventing accidents and hazards.
3. **Efficiency** – Through design and optimization, engineers help organizations operate more efficiently, saving time and resources.
4. **Compliance** – Engineering services ensure that projects comply with local regulations, standards, and environmental requirements.
5. **Sustainability** – Engineers play a key role in creating sustainable solutions that minimize environmental impact and promote long-term ecological balance.

14.1.3 Summary

In summary, Engineering services are diverse and essential in creating and maintaining the modern world, whether through designing infrastructure, managing construction, or innovating new technologies. Each service requires expertise and specialized knowledge to ensure success and safety in various industries.

14.2 GAS INSTALLATIONS SERVICES

Services on gas installations involve a range of activities aimed at ensuring that gas systems are safe, efficient, and compliant with regulations. These services are critical in residential, commercial, and industrial settings, where gas is used for heating, cooking, power generation, and other applications.

14.2.1 Comprehensive Gas System Services and Safety Solutions

1 Gas System Installation

- **Purpose** – Installing new gas systems, including pipes, meters, valves, and appliances, to ensure a safe and efficient supply of gas for various purposes.

Examples

- ***Residential Gas Heating*** – Installing a central gas heating system in a home, including gas boilers, radiators, and thermostats.
- ***Commercial Kitchen Setup*** – Setting up gas lines and connections for commercial-grade stoves, ovens, and grills in a restaurant.

2 Gas Leak Detection and Repair

- **Purpose** – Identifying and repairing leaks in gas systems to prevent potential hazards such as explosions, fires, and health risks.

Examples

- ***Electronic Leak Detection*** – Using specialized electronic detectors to identify leaks in underground pipelines or concealed gas lines.
- ***Pipeline Repair*** – Repairing a damaged gas line in an industrial plant to stop gas from escaping and posing a safety risk.

3 Gas Safety Inspection and Certification

- **Purpose** – Conducting thorough inspections of gas installations to ensure compliance with safety standards and local regulations. These inspections often result in a safety certification for the property or facility.

Examples

- *Boiler Inspection* – Checking a gas boiler's safety devices, flue, pressure settings, and ventilation to ensure it's operating safely.
- *Annual Gas Safety Check* – Performing a comprehensive annual safety inspection in residential apartments to identify any potential risks or issues.

4 Routine Maintenance and Servicing

- **Purpose** – Regular maintenance and servicing of gas appliances and systems to extend their lifespan, improve efficiency, and prevent malfunctions.

Examples

- *Gas Heater Servicing* – Cleaning and inspecting the burners, pilot lights, and heat exchangers of a gas heater to ensure proper functionality.
- *Commercial Boiler Maintenance* – Routine servicing of industrial gas boilers, including descaling, checking combustion efficiency, and testing safety valves.

5 Gas Appliance Installation and Commissioning

- **Purpose** – Installing new gas appliances and commissioning them to ensure they function correctly. This includes setting up appliances, checking connections, and verifying performance.

Examples

- *Gas Stove Installation* – Installing a gas stove in a home kitchen, ensuring the gas connection is secure, and the appliance is functioning properly.
- *Industrial Gas Furnace Setup* – Setting up and commissioning a high-efficiency gas furnace in a manufacturing plant, including adjusting airflow and combustion rates.

6 Pressure Testing and Tightness Testing

- **Purpose** – Conducting pressure and tightness tests on gas pipelines to ensure there are no leaks or weak points. These tests verify the integrity of the system and ensure safety.

Examples

- *Pressure Test of New Gas Lines* – Conducting a pressure test on newly installed gas pipelines in a residential development to verify they can handle the expected load.
- *Tightness Testing of Industrial Gas Systems* – Performing a tightness test on industrial gas systems to detect any small leaks or pressure drops over time.

7 Pipework Installation and Replacement

- **Purpose** – Installing or replacing gas pipes to accommodate new installations or upgrade old systems. This is crucial for ensuring the safe and efficient transport of gas from the supply source to the end-user.

Examples

- ***Pipe Upgrading for New Capacity*** – Replacing old gas pipes in a commercial building to accommodate a higher-capacity gas heating system.
- ***New Gas Line Installation*** – Installing additional gas lines in an industrial facility to power new equipment or machinery.

8 Ventilation and Flue Installation

1 **Purpose** – Installing proper ventilation and flue systems to safely vent exhaust gases produced by gas appliances, ensuring indoor air quality and safety.

Examples

- ***Chimney Flue Installation*** – Installing a chimney flue for a gas fireplace to ensure proper ventilation and prevent carbon monoxide buildup in a home.
- ***Industrial Ventilation Setup*** – Setting up a comprehensive exhaust system in an industrial kitchen to manage fumes and gases from multiple high-output gas appliances.

9 Gas Meter Installation and Replacement

- **Purpose** – Installing, replacing, or relocating gas meters to measure gas consumption accurately for billing or monitoring purposes.

Examples

- ***New Gas Meter Setup*** – Installing a new gas meter in a residential home to connect the household to the main gas supply.
- ***Industrial Meter Replacement*** – Replacing an outdated gas meter in a factory with a digital, high-precision model to improve energy tracking.

10 Combustion Analysis and Optimization

- **Purpose** – Analysing and optimizing the combustion process of gas-powered appliances to improve efficiency, reduce fuel consumption, and minimize emissions.

Examples

- ***Boiler Combustion Tuning*** – Conducting a combustion analysis on a gas boiler to optimize the air-fuel ratio for efficient burning and lower emissions.

- *Burner Adjustment in Industrial Furnace* – Fine-tuning the gas burner in an industrial furnace to achieve the desired temperature while minimizing gas usage.

11 Emergency Services and Troubleshooting

- **Purpose** – Providing rapid response to emergencies such as gas leaks, appliance malfunctions, or system failures. This includes troubleshooting and immediate repair to mitigate risks.

Examples

- *Gas Leak Emergency Response* – Responding to a reported gas leak in a residential neighborhood, identifying the source, and shutting off the supply until repairs are completed.
- *Appliance Breakdown Service* – Troubleshooting a commercial gas oven that has malfunctioned, repairing the fault, and restoring functionality.

12 System Upgrades and Retrofits

- **Purpose** – Upgrading or retrofitting existing gas systems to improve efficiency, safety, or to comply with new regulations. This can include replacing outdated equipment or enhancing system components.

Examples

- *Smart Thermostat Installation* – Upgrading a residential heating system with a smart thermostat for better control and energy efficiency.
- *Industrial Gas Burner Retrofit* – Retrofitting an industrial gas burner with a low-NOx (Nitrogen Oxide) burner to reduce emissions and meet environmental regulations.

13 Carbon Monoxide Monitoring and Alarm Installation

- **Purpose** – Installing carbon monoxide detectors and alarms to alert occupants if unsafe levels of carbon monoxide are detected, providing an additional layer of safety.

Examples

- *Home Carbon Monoxide Detectors* – Installing CO detectors near gas appliances in a home to monitor indoor air quality.
- *Commercial Detector System* – Setting up a carbon monoxide monitoring system in a commercial building's boiler room, complete with automated shut-off features for gas supply.

14.2.2 Key Takeaways on Gas Installation Services

- **Safety** – Ensuring the safety of gas installations is paramount, as gas leaks or malfunctions can lead to severe consequences, including explosions, fires, and carbon monoxide poisoning.
- **Compliance** – All gas services must adhere to strict industry standards and local regulations to ensure legal compliance and safe operation.
- **Efficiency** – Regular maintenance, upgrades, and optimization help improve the efficiency of gas systems, reducing energy costs and emissions.
- **Expertise** – Skilled technicians and certified professionals are required to handle gas installations due to the technical and safety challenges involved.

14.2.3 Summary

In summary, gas installation services play a vital role in residential, commercial, and industrial settings, ensuring that gas systems are functional, safe, and efficient for everyday use.

14.3 ELECTRICAL INSTALLATION SERVICES

Services on electrical installations cover a broad range of tasks to ensure that electrical systems are safe, reliable, efficient, and compliant with standards. These services are necessary in residential, commercial, and industrial environments to power equipment, lighting, and appliances safely and efficiently.

14.3.1 Common Services Carried Out on Electrical Installations

1 Electrical System Installation

- **Purpose** – Setting up new electrical systems, including wiring, panels, outlets, switches, and equipment. This includes planning and installing electrical components to ensure a consistent and safe power supply.

Examples

- ***Residential Wiring*** – Installing electrical wiring in a new home, connecting lighting, power outlets, and electrical panels.
- ***Commercial Office Setup*** – Installing lighting systems, network cabling, and power distribution for a new commercial office space.

2 Inspection and Testing of Electrical Systems

- **Purpose** – Conducting thorough inspections and tests to verify the safety, functionality, and compliance of electrical installations. This helps identify potential hazards or areas that need improvement.

Examples

- *Periodic Inspection* – Conducting a periodic inspection in a residential property to ensure the electrical system is still up to code and free of hazards.
- *PAT Testing (Portable Appliance Testing)* – Testing electrical appliances in a workplace to ensure they are safe for use, including checking cables, plugs, and internal connections.

3 Troubleshooting and Fault Diagnosis

- **Purpose** – Identifying and diagnosing faults in electrical systems. This service involves finding and fixing electrical problems, such as short circuits, electrical faults, or malfunctioning components.

Examples

- *Circuit Breaker Issues* – Diagnosing a tripping circuit breaker in a household and identifying the faulty appliance or overload causing the problem.
- *Industrial Machinery Faults* – Troubleshooting electrical faults in a piece of industrial machinery that is not functioning correctly, repairing wiring or control systems.

4 Electrical Maintenance and Repair

- **Purpose** – Performing routine maintenance and repair on electrical installations to ensure continuous operation and to prevent unexpected failures. This includes repairing damaged components and upgrading parts.

Examples

- *Fuse Replacement:* Replacing blown fuses in an electrical panel to restore power to a section of a building.
- *Industrial Motor Maintenance* – Regularly inspecting and servicing industrial motors to ensure they run efficiently and without interruption.

5 Upgrades and Retrofit Services

- **Purpose** – Upgrading or retrofitting existing electrical systems to improve efficiency, accommodate increased power needs, or comply with updated safety standards.

Examples

- o ***Lighting Upgrade*** – Replacing traditional lighting with energy-efficient LED lighting in a commercial building to reduce energy consumption.
- o ***Electrical Panel Upgrade*** – Upgrading an old electrical panel in a home to a modern circuit breaker system that can handle higher electrical loads.

6 Installation of Electrical Appliances and Equipment

- **Purpose** – Installing and connecting electrical appliances, machinery, or equipment to existing electrical systems. This service includes ensuring that installations are safe, functional, and efficient.

Examples

- o ***Kitchen Appliance Setup*** – Installing kitchen appliances like ovens, dishwashers, and microwaves in a residential kitchen, connecting them to the power supply.
- o ***HVAC Installation*** – Setting up and connecting HVAC systems in a commercial building, ensuring proper electrical wiring for heating, ventilation, and air conditioning units.

7 Circuit Design and Load Calculations

- **Purpose** – Designing electrical circuits and performing load calculations to ensure that electrical systems can handle the intended power consumption without overloading or causing failures.

Examples

- o ***Home Circuit Design*** – Designing the electrical layout for a home renovation, including the placement of outlets, switches, and lighting to ensure balanced loads.
- o ***Industrial Power Distribution*** – Designing a power distribution system for a factory to accommodate multiple machines and heavy equipment, ensuring each circuit can handle the load.

8 Energy Efficiency Audits and Optimization

- **Purpose** – Conducting energy efficiency audits to identify ways to reduce energy consumption, improve power efficiency, and minimize costs. This can involve upgrading or optimizing electrical systems.

Examples

- o ***Energy Audit in Office Buildings*** – Performing an energy audit in an office to recommend LED lighting, power-saving devices, and energy-efficient appliances.

- *Industrial Efficiency Improvements* – Implementing variable frequency drives (VFDs) for industrial motors to reduce energy use in a factory's production line.

9 Installation of Safety Systems

- **Purpose** – Installing and maintaining safety systems such as smoke detectors, fire alarms, emergency lighting, and surge protection to enhance safety and prevent accidents.

Examples

- *Fire Alarm System Installation* – Installing a comprehensive fire alarm system in a commercial building, including smoke detectors and manual pull stations.
- *Surge Protection Setup* – Setting up whole-house surge protectors to protect sensitive electronics from power surges.

10 Cable Management and Conduit Installation

- **Purpose** – Installing and organizing cables and conduits to protect electrical wiring and make maintenance easier. This service is vital for safety and aesthetics.

Examples

- *Commercial Cable Management* – Running cables in a commercial office space through conduits and cable trays to maintain organization and safety.
- *Concealed Wiring:* Installing concealed wiring in a residential home renovation to maintain a clean look while ensuring safety.

11 Renewable Energy Systems Installation

- **Purpose** – Installing renewable energy systems like solar panels, wind turbines, or battery storage systems to provide sustainable energy solutions.

Examples

- *Solar Panel Installation* – Setting up solar panels on the roof of a house, connecting them to the electrical grid or a battery backup system.
- *Wind Turbine Setup* – Installing small wind turbines on a commercial property to generate renewable electricity.

12 Emergency Services and Power Restoration

- **Purpose** – Providing emergency services for power outages, electrical faults, or safety incidents. This includes quick diagnosis and restoration of power to critical systems.

Examples

- *Power Outage Response* – Responding to a power outage in a neighborhood and restoring power by repairing damaged cables or transformers.
- *Industrial Emergency Repair* – Fixing a sudden electrical failure in an industrial facility that disrupts production, ensuring a swift return to operations.

13 Grounding and Bonding Services

- **Purpose** – Ensuring that electrical systems are properly grounded to prevent electrical shocks and reduce the risk of electrical fires. Bonding helps ensure that all metallic components are at the same electrical potential.

Examples

- *Residential Grounding Check* – Inspecting a home's grounding system to ensure it meets code requirements and replacing inadequate grounding rods.
- *Bonding in Industrial Sites* – Bonding metal structures, pipes, and equipment in a manufacturing facility to eliminate differences in electrical potential.

14.3.2 Importance of Electrical Installation Services

1. **Safety** – Proper installation and maintenance are essential to avoid electrical hazards like shocks, fires, and equipment malfunctions.
2. **Compliance** – Electrical services must adhere to strict building codes, regulations, and safety standards to avoid fines and ensure insurance coverage.
3. **Efficiency** – Well-maintained electrical systems operate more efficiently, reducing energy consumption and operational costs.
4. **Reliability** – Regular maintenance prevents unexpected breakdowns, ensuring continuous power supply for critical applications.
5. **Sustainability** – Services like energy audits and renewable energy installations promote sustainable energy use, reducing carbon footprints.

14.3.3 Summary

In summary, electrical installation, services are diverse and integral to modern living, supporting the safe, efficient, and reliable use of electricity in all sectors.

14.4 STEAM INSTALLATIONS SERVICES

Services on steam installations involve a variety of tasks to ensure the safe, efficient, and reliable operation of steam systems. Steam is widely used in industrial and commercial settings for heating, power generation, sterilization, and other processes. Proper maintenance and installation of steam systems are critical to maximizing efficiency and minimizing hazards.

14.4.1 Complete Steam Boiler and System Solutions for Efficiency and Safety

1 Steam Boiler Installation and Sctup

- **Purpose** – Installing steam boilers to generate steam for industrial, commercial, or residential applications. This involves setting up the boiler, connecting pipes, and ensuring proper fuel and water supply.

Examples

- o ***Industrial Boiler Installation*** – Setting up a high-capacity steam boiler in a manufacturing facility for use in production processes.
- o ***Commercial Boiler Setup*** – Installing a smaller steam boiler in a hotel to provide heating and hot water for the facility.

2 Steam System Design and Engineering

- **Purpose** – Designing and cngineering steam systems, including pipework, control systems, and auxiliary components, to meet the specific needs of the facility. This includes load calculations, sizing, and layout.

Examples

- o ***Steam Distribution System Design*** – Creating a steam distribution network for a factory, ensuring steam is delivered efficiently to various production lines.
- o ***Boiler Room Layout*** – Designing the layout of a boiler room in a hospital to ensure accessibility, safety, and compliance with regulations.

3 Routine Maintenance and Servicing of Boilers

- **Purpose** – Performing regular maintenance and servicing on steam boilers to ensure they operate safely and efficiently. This includes inspections, cleaning, and testing critical components.

Examples

- o ***Annual Boiler Inspection*** – Conducting an annual inspection of a steam boiler to check for leaks, corrosion, scaling, and proper combustion.

- ***Descaling*** – Removing scale build-up from a steam boiler's tubes to improve efficiency and prevent overheating.

4 Steam Trap Testing and Replacement

- **Purpose** – Inspecting, testing, and replacing steam traps to ensure proper condensate drainage and to prevent steam loss. Steam traps are essential for maintaining system efficiency and avoiding waste.

Examples

- ***Trap Inspection*** – Performing a diagnostic check on steam traps in a manufacturing plant to ensure they are functioning correctly.
- ***Steam Trap Replacement*** – Replacing faulty steam traps in a processing plant to stop leaks and save energy.

5 Pipework Installation and Insulation

- **Purpose** – Installing and insulating steam pipes to ensure efficient steam delivery and minimize heat loss. Proper insulation reduces energy consumption and prevents burns from hot surfaces.

Examples

- ***Pipe Installation*** – Installing new steam pipes in a chemical processing plant to handle increased steam demand.
- ***Insulation Upgrades*** – Re-insulating old steam pipes in a food processing facility to reduce energy costs and improve safety.

6 Control System Installation and Calibration

- **Purpose** – Installing, configuring, and calibrating control systems to regulate steam pressure, temperature, and flow rates. This includes setting up sensors, controllers, and automation systems.

Examples

- ***Steam Pressure Controller Installation*** – Setting up pressure regulators in a power plant to maintain consistent steam pressure for turbines.
- ***Automated Monitoring System*** – Installing an automated monitoring system in a textile mill to control steam flow based on production needs.

7 Safety Valve Testing and Certification

- **Purpose** – Testing and certifying safety valves to ensure they function correctly and provide pressure relief in case of over-pressurization. Safety valves are crucial for preventing accidents.

Examples

- o *Pop Test* – Performing a "pop test" on a steam boiler's safety valve to ensure it opens at the set pressure.
- o *Certification for Industrial Boilers* – Inspecting and certifying safety valves in a pharmaceutical plant to meet local safety codes.

8 Condensate Recovery System Installation and Maintenance

- **Purpose** – Installing and maintaining condensate recovery systems to capture and reuse condensed steam, saving energy and reducing water consumption.

Examples

- o *Condensate Pump Setup* – Installing a condensate pump in a hospital's steam system to return condensate to the boiler.
- o *Heat Exchanger Maintenance* – Servicing heat exchangers in a refinery's steam system to improve condensate recovery and heat transfer efficiency.

9 Water Treatment and Boiler Chemistry Management

- **Purpose** – Managing boiler water chemistry through treatment and monitoring to prevent corrosion, scaling, and fouling in steam systems. Proper water quality is vital for efficient boiler operation.

Examples

- o *Chemical Dosing* – Adding corrosion inhibitors and scale prevention chemicals to a steam boiler's feedwater in a power plant.
- o *Water Testing* – Regularly testing the water in a food processing plant's steam boiler to ensure it meets the required chemical specifications.

10 Pressure Testing and Leak Detection

- **Purpose** – Conducting pressure tests on steam systems to identify potential leaks, weaknesses, or pressure drops. Early detection prevents costly repairs and safety hazards.

Examples

- *Hydrostatic Pressure Test* – Performing a hydrostatic pressure test on new steam pipes in a chemical factory to check for leaks.
- *Ultrasonic Leak Detection* – Using ultrasonic equipment to detect small steam leaks in a refinery's steam network.

11 Combustion Analysis and Boiler Efficiency Optimization

- **Purpose** – Conducting combustion analysis to ensure efficient fuel usage in steam boilers and to optimize the combustion process for reduced emissions and fuel costs.

Examples

- *Burner Tuning* – Adjusting the air-fuel ratio in a boiler's burner to achieve complete combustion and improve efficiency in a pulp and paper mill.
- *Flue Gas Analysis* – Analysing flue gases in a large industrial boiler to identify combustion efficiency and make necessary adjustments.

12 Steam System Upgrades and Retrofits

- **Purpose** – Upgrading or retrofitting steam systems to improve performance, enhance safety, or meet new regulations. This can involve replacing old equipment or adding new components.

Examples

- *Low-NOx Burner Retrofit* – Retrofitting a steam boiler with a low-NOx burner in a manufacturing plant to reduce emissions and comply with environmental standards.
- *Heat Recovery Steam Generator (HRSG)* – Adding a heat recovery steam generator to capture waste heat from industrial processes in a power plant.

13 Emergency Services and Breakdown Repairs

- **Purpose** – Providing emergency repair services for steam systems in case of unexpected failures, leaks, or malfunctions to minimize downtime and ensure safety.

Examples

- *Emergency Boiler Shutdown* – Shutting down a steam boiler in a chemical plant during a sudden pressure spike and repairing the underlying cause.
- *Pipe Burst Repair* – Quickly fixing a burst steam pipe in an industrial setting to restore operations and prevent further damage.

14 **Steam Turbine Installation and Maintenance**

- **Purpose** – Installing and maintaining steam turbines used for power generation or mechanical work in industrial facilities. Regular maintenance is necessary to keep turbines operating efficiently.

Examples

- *Turbine Installation* – Installing a steam turbine in a power generation plant, connecting it to the steam supply, and configuring control systems.
- *Turbine Overhaul* – Performing a major overhaul of a steam turbine in a chemical processing plant to inspect and replace worn components.

14.4.2 Key Takeaways on Steam Installation Services

1 **Safety** – Proper installation, maintenance, and servicing are critical to ensuring the safety of steam systems, as they operate under high pressure and temperature conditions.

2 **Efficiency** – Regular maintenance, including the replacement of faulty components and optimization of the system, helps to improve efficiency and reduce energy consumption.

3 **Reliability** – Reliable steam systems are essential for continuous operation in industrial and commercial settings, avoiding costly downtime.

4 **Sustainability** – Services like condensate recovery and combustion optimization contribute to sustainable operations by conserving energy and reducing environmental impact.

5 **Compliance** – Steam installations must adhere to strict safety and environmental standards to ensure legal compliance and safe operation.

14.4.3 Summary

Overall, services carried out on steam installations are crucial for industries that rely on steam for heating, power, or processing, ensuring that these systems are safe, efficient, and fit for purpose.

14.5 HOT AND COLD-WATER INSTALLATIONS

Services carried out on hot and cold-water installations involve a range of tasks to ensure the efficient and safe supply of water for residential, commercial, and industrial use. Proper installation and maintenance are essential for ensuring consistent water pressure, temperature control, and system reliability.

14.5.1 Breakdown of Common Services Related to Hot and Cold-Water Installations

1 Installation of Water Supply Systems

- **Purpose** – Setting up the water supply infrastructure, including pipes, pumps, and fixtures, to ensure the reliable distribution of both hot and cold water. This is a foundational step in any new building or renovation project.

Examples

- ***Residential Water Supply*** – Installing pipes for hot and cold-water distribution in a new home, connecting them to a main supply line and water heater.
- ***Commercial Water Distribution*** – Setting up a complex water distribution system in a hotel to supply cold and hot water to guest rooms, kitchens, and laundry areas.

2 Water Heater Installation and Maintenance

- **Purpose** – Installing and servicing water heaters, including traditional tank-style heaters, tankless heaters, and solar water heaters. Proper installation ensures energy efficiency and safety.

Examples

- ***Gas Water Heater Installation*** – Installing a gas water heater in a residential property, connecting it to the gas line, and ensuring proper venting.
- ***Tankless Water Heater Maintenance*** – Performing annual maintenance on a tankless water heater in a commercial kitchen to remove mineral build-up and maintain efficiency.

3 Pipe Installation and Replacement

- **Purpose**: Installing, repairing, or replacing pipes in a water system to ensure proper flow and prevent leaks. This includes selecting appropriate materials such as copper, PEX, or PVC.

Examples

- ***PEX Pipe Installation*** – Replacing old galvanized pipes in a house with modern PEX pipes to improve water quality and flow.
- ***Copper Pipe Installation*** – Installing copper pipes in a commercial property for hot water distribution due to their heat resistance and durability.

4 Leak Detection and Repair

- **Purpose** – Identifying and repairing leaks in hot and cold-water systems to prevent water loss, property damage, and increased utility costs. Leak detection can involve advanced tools like acoustic detectors or thermal imaging.

Examples

- *Leak Repair in Kitchen* – Fixing a leaking cold-water pipe under the kitchen sink in a residential property.
- *Commercial Leak Detection* – Using thermal imaging to identify a hot water leak behind walls in a commercial building without causing damage.

5 **Installation of Mixing Valves and Temperature Control Devices**

- **Purpose** – Installing and maintaining mixing valves to regulate the temperature of hot and cold water, ensuring consistent and safe water temperatures. This is especially important in commercial settings where scald protection is critical.
- **Examples**
  - *Thermostatic Mixing Valve Setup* – Installing a thermostatic mixing valve in a hospital to maintain safe temperatures in showers and sinks.
  - *Anti-Scald Device Installation* – Adding anti-scald devices to residential bathroom fixtures to prevent burns from hot water.

6 **Water Pressure Testing and Adjustment**

- **Purpose** – Testing and adjusting water pressure to ensure that it meets optimal standards for both hot and cold-water systems. Excessive pressure can cause leaks, while low pressure can reduce flow efficiency.

Examples

- *Pressure Regulator Installation* – Installing a pressure regulator in a residential property to maintain consistent water pressure.
- *Commercial Pressure Adjustment* – Adjusting water pressure in a multi-story office building to ensure proper flow to upper floors.

7 **Installation of Water Storage and Booster Systems**

- **Purpose** – Setting up water storage tanks and booster systems to ensure adequate water supply and pressure for larger buildings or areas with low municipal water pressure.

Examples

- *Residential Water Tank Installation* – Installing a cold-water storage tank in a home with unreliable municipal water supply to ensure consistent water availability.
- *Commercial Booster Pump Setup* – Installing a booster pump in a hotel to maintain consistent water pressure during peak usage periods.

8 Insulation of Hot Water Pipes

- **Purpose** – Insulating hot water pipes to reduce heat loss, improve energy efficiency, and prevent condensation on cold-water pipes. Insulation is crucial in areas with temperature extremes.

Examples

- ***Pipe Insulation in New Construction*** – Installing pipe insulation in a new residential property to improve energy efficiency and reduce utility bills.
- ***Commercial Hot Water System*** – Insulating the hot water pipes in a factory to maintain water temperature and reduce energy consumption.

9 Backflow Prevention Installation and Testing

- **Purpose** – Installing and testing backflow prevention devices to protect the potable water supply from contamination due to backflow. This is required for compliance with health and safety regulations.

Examples

- ***Backflow Preventer Installation*** – Installing a backflow preventer on the main water supply line of a commercial building to protect against contamination.
- ***Annual Testing*** – Performing annual backflow testing in an industrial facility to ensure compliance with local regulations.

10 Descaling and Flushing Hot Water Systems

- **Purpose** – Removing scale and sediment build-up from hot water systems to improve efficiency and prolong the lifespan of water heaters and pipes. Hard water can cause scaling, affecting performance.

Examples

- ***Tank Water Heater Descaling*** – Flushing a traditional tank-style water heater to remove mineral build-up and maintain heating efficiency.
- ***Tankless Heater Flushing*** – Using a vinegar solution to flush a tankless water heater in a restaurant to remove limescale.

11 Cold-water Filtration and Treatment

- **Purpose** – Installing and maintaining filtration systems to ensure the quality of cold water, removing impurities, sediments, or contaminants.

Examples

- ***Whole House Filtration System*** – Installing a whole-house filtration system in a residential property to remove chlorine and improve water quality.
- ***Point-of-Use Filters:*** Setting up under-sink water filters in an office breakroom to provide clean drinking water.

12 Hot and Cold-water System Balancing

- **Purpose** – Balancing hot and cold-water systems to ensure even distribution throughout a building. Balancing valves may be used to adjust flow rates and temperatures for consistency.

Examples

- ***System Balancing in Hospitals*** – Balancing hot and cold-water supply in a hospital to ensure even temperatures and pressure in all patient areas.
- ***Balancing Valves Setup*** – Installing balancing valves in a multi-story apartment complex to ensure consistent water temperature at all levels.

13 Emergency Repairs and Replacement Services

- **Purpose** – Providing emergency repair and replacement services for sudden failures in hot and cold-water systems to prevent damage, ensure safety, and restore functionality.

Examples

- ***Burst Pipe Repair*** – Quickly fixing a burst cold-water pipe in a commercial building to prevent flooding.
- ***Hot Water Heater Replacement*** – Replacing a failed water heater in a residential property on an emergency basis to restore hot water supply.

14 Integration of Smart Water Management Systems

- **Purpose** – Installing smart water management systems for monitoring and controlling hot and cold-water installations remotely. These systems help optimize water usage and detect leaks early.

Examples

- ***Smart Thermostats*** – Installing a smart thermostat that controls hot water temperature based on usage patterns in a residential home.
- ***Leak Detection System:*** Setting up smart sensors in a commercial building to monitor for leaks and water usage anomalies.

15 Fixture Installation and Maintenance

- **Purpose** – Installing, replacing, and maintaining water fixtures, such as faucets, showers, sinks, and toilets, to ensure proper operation and prevent leaks.

Examples

- ***Faucet Replacement*** – Replacing old kitchen faucets with new ones in a home renovation.
- ***Commercial Fixture Maintenance*** – Servicing and repairing fixtures in a hotel bathroom to prevent leaks and maintain aesthetics.

14.5.2 Key Takeaways on Hot and Cold-Water Installation Services

1. **Safety** – Proper installation and regular maintenance are essential for preventing leaks, water damage, and temperature hazards, ensuring a safe water supply.
2. **Efficiency** – Services like pipe insulation, system balancing, and leak detection optimize energy and water use, reducing costs and improving performance.
3. **Reliability** – Routine maintenance and inspections ensure continuous water supply, avoiding unexpected failures that can disrupt daily activities.
4. **Compliance** – Services must meet local codes and standards, particularly concerning backflow prevention and water quality, to protect public health.
5. **Sustainability** – Smart water management and energy-efficient practices promote sustainable use of resources, helping to reduce environmental impact.

14.5.3 Summary

In summary, services carried out on hot and cold-water installations are crucial for ensuring a safe, efficient, and reliable supply of water in any building, supporting the comfort and well-being of occupants.

14.6 HYDRAULIC AND PNEUMATIC SYSTEMS

Hydraulic and pneumatic systems are essential technologies in various industrial applications, relying on the power of fluids or compressed air to generate mechanical force. These systems are commonly used in machinery and equipment for automation, manufacturing, construction, and many other fields.

14.6.1 Hydraulic Systems

Hydraulic systems use incompressible fluids (usually oil) to transmit power. They operate based on Pascal's Law, which states that pressure applied to a confined fluid is transmitted equally in all directions. This makes hydraulic systems ideal for applications requiring high force and precise control.

14.6.1.1 Key Components of Hydraulic Systems

1 **Hydraulic Pump** – Generates fluid flow and pressure.

2 **Hydraulic Fluid** – Usually oil, transmits power throughout the system.

3 **Actuators (Cylinders and Motors)** – Convert hydraulic energy into mechanical movement or force.

4 **Valves** – Control the direction, pressure, and flow of hydraulic fluid.

5 **Reservoir** – Stores hydraulic fluid.

6 **Piping** – Transports hydraulic fluid to different parts of the system.

14.6.1.2 Examples of Hydraulic Systems

1 Construction Equipment (Excavators and Loaders)

- ***Function*** – Hydraulic systems are commonly found in heavy construction equipment, such as excavators and loaders, to control the movement of arms, booms, and buckets.
- ***Application*** – In an excavator, hydraulic cylinders lift the arm and control the digging motion, while hydraulic motors rotate the turret.
- ***Advantage*** – Provides high force, precise control, and the ability to handle heavy loads.

2 Hydraulic Presses

- ***Function*** – These presses use hydraulic cylinders to apply force, often used for molding, forging, or stamping metal parts.
- ***Application*** – In the automotive industry, hydraulic presses form metal panels and components with significant force.
- ***Advantage*** – Consistent pressure, adjustable force, and precise control make them suitable for repetitive industrial tasks.

3 Aircraft Landing Gear

- ***Function*** – In aircraft, hydraulic systems control landing gear retraction and extension, as well as other functions like braking and steering.
- ***Application*** – Hydraulic actuators raise and lower the landing gear smoothly and retract it into the aircraft's fuselage.

- *Advantage* – Provides reliable, high-power output necessary for critical flight control systems.

14.6.2 Pneumatic Systems

Pneumatic systems use compressed air or gas to transmit power. They are generally simpler and cleaner than hydraulic systems, making them suitable for applications that do not require extremely high force. Pneumatics are ideal for tasks requiring fast, repetitive movements.

14.6.2.1 Key Components of Pneumatic Systems

- **Compressor** – Generates compressed air.
- **Air Reservoir** – Stores compressed air.
- **Actuators (Cylinders and Motors)** – Convert compressed air into mechanical movement or force.
- **Valves** – Control airflow, direction, and pressure.
- **Tubing** – Carries compressed air to different parts of the system.
- **Filters and Lubricators** – Ensure clean air supply and reduce friction.

14.6.2.2 Examples of Pneumatic Systems

1 **Manufacturing Automation (Robotic Arms and Conveyors)**

- *Function* – Pneumatic systems are widely used in automated manufacturing processes to power robotic arms, conveyors, and pick-and-place machines.
- *Application* – Pneumatic cylinders move robotic arms quickly and precisely for assembly, packaging, or handling materials.
- *Advantage* – Fast response time, simplicity, and lower maintenance costs compared to hydraulics.

2 **Pneumatic Braking Systems in Trucks and Buses**

- *Function* – Pneumatic systems are commonly used in heavy vehicles for braking. Compressed air applies force to the brake pads, slowing or stopping the vehicle.
- *Application* – In a truck, pressing the brake pedal releases compressed air to engage the brake mechanism on each wheel.
- *Advantage* – Simple, reliable, and failsafe—brakes remain engaged if air pressure is lost.

3 Dentist Drills

- *Function* – In dental clinics, pneumatic systems power high-speed dental drills.
- *Application* – Compressed air rotates the drill bit at high speed, allowing dentists to perform precise procedures.
- *Advantage* – Lightweight, easy to control, and quieter than electric drills, making them ideal for a clinical environment.

14.6.3 Table of Key Differences Between Hydraulic and Pneumatic Systems

Feature	Hydraulic Systems	Pneumatic Systems
Medium	Incompressible fluid (usually oil)	Compressed air or gas
Force	High force (can handle heavy loads)	Moderate force (suitable for lighter loads)
Precision	High precision	Moderate precision
Speed	Slower, but powerful	Faster, but less forceful
Cost	Generally, more expensive	Usually, cheaper
Maintenance	Requires frequent maintenance and fluid checks	Low maintenance, easier to clean
Applications	Heavy machinery, aircraft, manufacturing	Automation, packaging, small machinery

14.6.4 Applications of Hydraulic and Pneumatic Systems in Industry

1 Automotive Industry

- *Hydraulics* – Used in car jacks, brake systems, and hydraulic presses for metal shaping.
- *Pneumatics* – Powering robotic arms for assembly, tire inflation systems, and air brakes in commercial vehicles.

2 Manufacturing and Packaging

- *Hydraulics* – Utilized in injection molding machines, hydraulic presses, and metal forming equipment.
- *Pneumatics* – Conveyor systems, pneumatic presses, and automatic packaging machines for bottling or labeling.

3 **Mining and Construction**

- ***Hydraulics*** – Hydraulic excavators, rock drills, bulldozers, and cranes rely on hydraulic power for digging, lifting, and heavy construction.
- ***Pneumatics*** – Pneumatic drills, air hammers, and other tools used for drilling, breaking, and handling lighter construction tasks.

4 **Aerospace**

- ***Hydraulics*** – Controls critical components such as landing gear, flight control surfaces, and cargo doors.
- ***Pneumatics*** – Used in aircraft to operate non-critical components like air-driven pumps, pneumatic tools, and cabin pressurization.

5 **Healthcare**

- ***Hydraulics*** – Found in hospital beds, patient lifts, and specialized medical equipment requiring fine control and heavy lifting.
- ***Pneumatics*** – Utilized in ventilators, pneumatic syringes, and dental chairs for precise, clean, and reliable operation.

14.6.5 Advantages and Limitations of Hydraulic and Pneumatic Systems

- **Advantages of Hydraulic Systems**

1 High force output, suitable for heavy-duty applications.

2 Precision and smooth control over movements.

3 Capable of holding loads without additional power (hydraulic lock).

- **Limitations of Hydraulic Systems**

1 More complex and costly to install and maintain.

2 Prone to fluid leaks, which can cause contamination.

3 Slower response compared to pneumatic systems.

- **Advantages of Pneumatic Systems**

1 Fast and lightweight, making them ideal for repetitive tasks.

2 Cleaner operation since they use air, which is environmentally friendly.

3 Lower initial and maintenance costs.

- **Limitations of Pneumatic Systems**

1 Lower force output, not suitable for heavy loads.

2 Less precise than hydraulics for complex tasks.

3 Compressed air can be noisy and requires proper filtration to avoid moisture buildup.

14.6.6 Summary

In summary, hydraulic and pneumatic systems are versatile and essential for powering machinery in a wide range of industries. The choice between hydraulic and pneumatic systems depends on the specific needs of the application—hydraulics for high-force, precision, and heavy-duty tasks, and pneumatics for speed, simplicity, and lighter applications. Understanding the strengths and limitations of each system is crucial for optimizing efficiency, performance, and safety in industrial environments.

CHAPTER FIFTEEN

15.0 POWER TRANSMISSION SYSTEMS: FLUID DYNAMICS, FAULT DIAGNOSIS, SAFETY STANDARDS, AND EQUIPMENT CHARACTERISTICS

15.1 POWER TRANSMISSION DEVICES

When it comes to installing and maintaining basic transmission devices like belts, chains, ropes, and toothed gears, it is essential to adhere to specific safety and installation requirements tailored to each device. Each transmission type has its unique characteristics, strengths, and limitations. Below is a comparison of installation and safety requirements for these four common mechanical power transmission systems:

15.1.1 Belts

Belts are flexible loops that transfer power between rotating shafts. They are typically made from materials like rubber, polyurethane, or leather.

15.1.1.1 Installation Requirements

- **Alignment** – Ensure proper alignment between the driving and driven pulleys to prevent excessive wear and avoid belt slippage.
- **Tension** – Proper tension is crucial for optimal performance. Over-tensioning can damage bearings, while under-tensioning may cause slippage and reduce power transmission.
- **Pulley Size** – Use the correct pulley size to match the belt's specifications for speed and torque requirements.
- **Guarding** – Install safety guards to prevent accidental contact with moving belts and pulleys.
- **Cleanliness** – Keep the belt and pulley surfaces clean and free of debris to prevent contamination, slippage, and premature wear.

15.1.1.2 Safety Requirements

- **Guarding and Shields –** Provide proper guarding to prevent accidental contact with the moving belt and reduce the risk of clothing, hair, or limbs being caught.

- **Inspection** – Regularly inspect belts for signs of wear, fraying, or cracks, and replace them as needed.
- **Lubrication** – Some belt types may require lubrication, but it should be done with caution to avoid contamination and slipping.
- **Training** – Train personnel on safe belt installation, adjustment, and removal procedures, emphasizing lockout/tagout (LOTO) practices during maintenance.

15.1.2 Chains

Chains are robust transmission devices made of linked segments that can handle heavy loads. They are often used in applications requiring more power transmission than belts.

15.1.2.1 Installation Requirements

- **Alignment** – Ensure precise alignment between sprockets to prevent premature chain wear or breakage.
- **Tension** – Adjust chain tension to prevent excessive slack while allowing a small amount of free movement.
- **Lubrication** – Chains require regular lubrication to reduce friction, wear, and rust.
- **Sprocket Selection** – Choose the appropriate sprocket size and type to match the chain's pitch and strength requirements.
- **Guarding** – Install guards around exposed chain drives to protect against accidental contact.

15.1.2.2 Safety Requirements

- **Guarding** – Ensure complete guarding around chains to prevent accidents, especially in industrial settings where high-speed chain drives are in use.
- **Maintenance** – Regularly inspect chains for signs of wear, deformation, elongation, or damage. Replace worn chains immediately to prevent breakage.
- **Noise** – Chains can be noisy; consider using chain covers or dampers to reduce noise and vibrations.
- **Training** – Educate workers on the safe handling of chains, including tensioning, lubricating, and replacing them.

15.1.3 Ropes

Ropes (often wire ropes or synthetic ropes) are used for lifting, pulling, or transmitting motion in a variety of applications. They are more commonly used in hoisting and crane operations.

15.1.3.1 Installation Requirements

- **Tensioning** – Properly tension the rope to avoid sagging or excessive strain that can lead to failure.
- **Pulley Compatibility** – Use pulleys (sheaves) that are compatible with the rope's diameter and material to ensure proper fit and prevent wear.
- **Anchoring** – Secure rope ends with appropriate fittings (like clamps or splices) to prevent slippage.
- **Inspection** – Conduct pre-installation checks to identify any damage or defects in the rope.

15.1.3.2 Safety Requirements

- **Load Ratings** – Ensure that the rope's load rating matches the application's requirements. Overloading can lead to rope failure.
- **Inspection** – Regularly inspect ropes for signs of fraying, kinking, corrosion, or wear. Replace damaged or worn ropes immediately.
- **Environmental Considerations** – Protect ropes from environmental factors like moisture, chemicals, and UV light, which can degrade certain materials.
- **Personal Protective Equipment (PPE)** – Use appropriate PPE (like gloves) to handle ropes, particularly wire ropes, to prevent injuries from frayed wires.

15.1.4 Toothed Gears

Gears are rigid, toothed mechanical components that transmit motion and torque between shafts. They are used in high-torque applications requiring precise movement.

15.1.4.1 Installation Requirements

- **Alignment** – Ensure gears are properly aligned to avoid uneven load distribution and reduce noise and wear.
- **Lubrication** – Gears require proper lubrication to minimize friction and wear, and to dissipate heat generated during operation.

- **Clearance** – Maintain appropriate backlash (clearance between gear teeth) to ensure smooth meshing and avoid tooth damage.
- **Mounting** – Secure gears firmly to shafts using keys, splines, or other securing methods to prevent slipping or shifting.

15.1.4.2 Safety Requirements

- **Guarding** – Install guards to cover exposed gears, especially in high-speed or heavy-duty applications, to protect operators from accidental contact.
- **Inspection** – Regularly inspect gears for wear, broken teeth, cracks, or misalignment. Replace damaged gears promptly to prevent mechanical failure.
- **Lubrication Management** – Use proper lubrication techniques to avoid contamination, overheating, or gear seizing.
- **Noise Reduction** – Consider using sound-dampening enclosures or materials if gear noise is a concern, especially in quiet environments.

15.1.4.3 Table of Comparison of Installation and Safety Requirements

Feature	**Belts**	**Chains**	**Ropes**	**Toothed Gears**
Alignment	High priority for reducing wear and slippage	Critical to prevent wear or breakage	Important to ensure smooth load distribution	Essential to avoid noise and uneven wear
Tension	Proper tensioning required	Needs appropriate tension to prevent slack	Requires tensioning to avoid sagging	Not applicable
Lubrication	Minimal, depends on type	Regular lubrication needed	May need lubrication for wire ropes	Continuous lubrication needed
Guarding	Essential to prevent contact with moving parts	Critical, especially in industrial settings	Necessary, especially with wire ropes	Essential, particularly for high-speed gears
Inspection	Check for fraying, wear, and cracks	Inspect for wear, elongation, and damage	Inspect for fraying, corrosion, and wear	Inspect for tooth wear, cracks, and misalignment

Feature	Belts	Chains	Ropes	Toothed Gears
Noise	Relatively quiet operation	Can be noisy; dampers may be needed	Quiet, depending on material	Can be noisy; requires damping or enclosures

15.1. 5 Summary

Each type of transmission device—belts, chains, ropes, and gears—has distinct installation and safety requirements. Proper alignment, tension, lubrication, guarding, and regular inspection are the key factors for safe and efficient operation. Selecting the right transmission device for the specific industrial task will ensure optimal performance, safety, and longevity of the machinery. Understanding the unique requirements of each transmission type is crucial for maintenance personnel, engineers, and operators working in industrial environments.

15.2 CHARACTERISTICS OF FLUID POWER TRANSMISSION - CONSTANT AND AVAILABLE SPEED DEVICES

Fluid power transmission, which uses fluids (liquid or gas) to transmit energy, is a crucial technology in various industrial applications. It is broadly categorized into hydraulic systems (using liquids like oil) and pneumatic systems (using gases like compressed air). These systems are known for their efficiency, control flexibility, and ability to handle heavy loads.

The two primary characteristics of fluid power transmission systems are their capacity to provide both constant speed and variable speed control.

15.2.1 Constant Speed Fluid Power Devices

Constant speed fluid power devices are designed to maintain a fixed output speed regardless of load variations. These devices are used in applications where consistent speed is crucial for the proper functioning of machinery.

15.2.1.1 Characteristics of Constant Speed Devices

- **Steady Output** – Constant speed devices provide consistent, predictable output. This makes them suitable for operations where maintaining a uniform speed is critical, such as conveyor systems or certain automated manufacturing processes.
- **Reliable Torque** – They deliver reliable torque, which is essential for applications needing precise and stable performance.

- **Less Complex Controls** – These systems often have simpler control mechanisms, which can reduce maintenance requirements and simplify operation.
- **Fixed Displacement Pumps/Motors** – Many hydraulic systems that operate at a constant speed use fixed displacement pumps or motors. These pumps deliver a set volume of fluid per revolution, leading to a constant output speed as long as the input remains consistent.
- **Efficiency** – Constant speed systems can be more energy-efficient in applications where the load is constant, minimizing energy waste and enhancing system longevity.

15.2.1.2 Examples of Constant Speed Devices

- **Hydraulic Pumps (Fixed Displacement)** – These pumps are designed to deliver a constant flow rate. They are ideal for systems that do not require variable speed control.
- **Hydraulic Motors** – When paired with a constant flow hydraulic pump, the hydraulic motor will run at a constant speed, regardless of load changes.
- **Air Compressors** – In pneumatic systems, some compressors are set to maintain a constant air pressure, leading to a consistent operational speed for connected tools.

15.2.2 Variable Speed Fluid Power Devices

Variable speed devices are designed to offer flexibility in controlling the speed and force applied to the load. They allow the operator to adjust the speed, making them ideal for applications that require versatility and fine control.

15.2.2.1 Characteristics of Variable Speed Devices

- **Adjustable Output** – These devices provide adjustable speeds to match the requirements of varying loads, making them suitable for applications like hoisting, lifting, or variable manufacturing tasks.
- **Precision Control** – They allow for precise control over both speed and force, which is crucial for sensitive operations like assembly lines, where the speed may need to adjust during different production phases.
- **Complex Control Systems** – Variable speed systems often require more complex control mechanisms like proportional valves, electronic controllers, or variable displacement pumps to manage changes in speed and force accurately.
- **Variable Displacement Pumps/Motors** – These are commonly used in hydraulic systems that need variable speed control. They can adjust the fluid flow, leading to a change in the output speed.

- **Energy Efficiency** – Although they have more complex controls, variable speed systems can be more energy-efficient in applications with fluctuating loads, as they only use the energy required for a specific task.

15.2.2.2 Examples of Variable Speed Devices

- **Variable Displacement Hydraulic Pumps** – These pumps can change the amount of fluid delivered per stroke, allowing the system to operate at different speeds without changing the input.
- **Proportional Valves** – In hydraulic systems, proportional valves regulate the flow of fluid to control speed and direction accurately.
- **Pneumatic Motors with Variable Air Pressure** – These motors adjust speed based on the amount of compressed air provided, allowing for precise speed adjustments in pneumatic applications.
- **Servo-Hydraulic Systems** – These are advanced hydraulic systems that use feedback sensors and servo controllers to deliver precise variable speeds.

15.2.2.3 Table of Comparison Between Constant and Variable Speed Fluid Power Devices

Aspect	Constant Speed Devices	Variable Speed Devices
Speed	Fixed, consistent speed regardless of load	Adjustable speed to match varying load requirements
Control Complexity	Simple control mechanisms	Requires more complex controls for precise adjustments
Torque Delivery	Reliable and consistent	Adjustable torque based on speed and load needs
Energy Efficiency	More efficient for steady-load applications	More efficient for variable-load applications
Applications	Suitable for uniform operations (e.g., conveyors)	Ideal for dynamic operations (e.g., variable-speed production)
Pump Types	Uses fixed displacement pumps/motors	Utilizes variable displacement pumps/motors
Maintenance	Easier to maintain due to simpler design	May require more frequent maintenance due to complexity

15.2.3 Fluid Power Transmission Examples

15.2.3.1 Hydraulic Systems

- **Excavators** – These heavy machines use variable-speed hydraulic systems for precise digging and lifting. The operator can control the speed of the arm movements, adjusting the hydraulic flow as needed for accuracy.
- **Injection Molding Machines** – These often use constant speed hydraulic pumps to maintain consistent pressure for injecting plastic into molds. The pump maintains a constant flow rate to ensure uniform product quality.
- **Hydraulic Presses** – Depending on the application, they can use constant or variable speed systems. A constant-speed press may be used for repetitive tasks, while a variable-speed system might be used for tasks requiring different force levels.

15.2.3.2 Pneumatic Systems

- **Assembly Line Tools**: Pneumatic tools often rely on variable-speed air motors, which adjust speed based on air pressure. This enables workers to fine-tune the tool's performance for different tasks.
- **Conveyor Belts**: Pneumatic systems can be designed to move at a constant speed, ensuring steady product movement, or be adjusted for variable speeds to match production demands.
- **Braking Systems**: Pneumatic brakes, especially in heavy vehicles, can vary the force applied depending on the pressure in the system, allowing for both constant and variable speed braking.

15.2.4 Summary

Fluid power transmission is a versatile and reliable way to transfer energy in industrial systems. Constant speed devices are suitable for applications where a uniform, steady output is needed, while variable speed devices excel in situations requiring precision, flexibility, and control. Understanding the characteristics of each type is essential for selecting the appropriate system based on the specific needs of the operation. Proper maintenance, safety, and operation are critical to ensuring the efficiency and longevity of fluid power systems.

15.3 DIAGNOSING AND REPAIRING FAULTS IN FLUID POWER TRANSMISSION SYSTEMS

Fluid power transmission systems, whether constant speed or variable speed, are susceptible to various faults that can affect performance, safety, and efficiency. Proper diagnosis involves identifying symptoms, locating the source of the problem, and implementing appropriate repairs.

15.3.1 Common Symptoms, Fault Identification, and Repair Techniques for Constant and Variable Speed Fluid Power Devices.

15.3.1.1 Identifying Symptoms in Fluid Power Systems

Common Symptoms

- **Loss of Pressure** – Decreased pressure output leading to insufficient power for hydraulic or pneumatic operations.
    - ***Example*** – A hydraulic press fails to exert the expected force, leading to defective parts or inconsistent results.
- **Slow or Erratic Movement** – Actuators (cylinders or motors) move slower than expected, hesitate, or behave erratically.
    - ***Example*** – A pneumatic drill that should run continuously exhibits intermittent speed fluctuations.
- **Overheating** – Excessive heat generation in the system components, leading to decreased efficiency or potential damage.
    - ***Example*** – A hydraulic pump gets unusually hot, causing deterioration of seals and fluid.
- **Unusual Noises** – Abnormal noises like knocking, grinding, whining, or squealing from components.
    - ***Example*** – A hydraulic motor emits a loud screeching sound during operation, indicating potential internal wear or cavitation.
- **Fluid Leaks** – Visible oil or air leaks around fittings, hoses, or seals, leading to loss of pressure and contamination.
    - ***Example*** – A hydraulic hose shows fluid seepage, causing a reduction in power and mess around the equipment.
- **Vibration** – Excessive vibrations or shaking in the equipment can indicate imbalances, misalignments, or worn components.
    - ***Example*** – A constant-speed pump vibrates heavily during operation, suggesting issues with alignment or internal damage.
- **Stuck or Unresponsive Components** – Actuators, valves, or motors do not respond to controls, indicating potential blockages or failures.

- ***Example*** – A valve in a variable-speed hydraulic system does not change position when the controller is adjusted.

15.3.1.2 Locating Faults in Fluid Power Systems

Common Fault Locations

- **Pump Issues**
  - ***Fault Symptoms:*** Loss of pressure, overheating, slow or erratic movement.
  - ***Diagnostic Steps*** – Check for worn-out pump components, cavitation, misalignment, or air leaks in suction lines.
  - ***Tools*** – Pressure gauges, thermal sensors, and flow meters can help identify issues with the pump.
- **Valve Failures**
  - ***Fault Symptoms*** – Erratic movement, stuck components, loss of pressure.
  - ***Diagnostic Steps*** – Inspect control valves for blockages, dirt, or internal wear. Use diagnostic tools like flow testers to ensure proper flow and pressure.
  - ***Tools*** – Flow meters and diagnostic pressure gauges can help pinpoint malfunctioning valves.
- **Hose and Seal Leaks**
  - ***Fault Symptoms*** – Fluid leaks, loss of pressure, slow movements.
  - ***Diagnostic Steps*** – Visually inspect hoses, connections, and seals for cracks, wear, or loosening. Perform a bubble test for air leaks in pneumatic systems.
  - ***Tools*** – Visual inspection, soap solution for air leaks, and UV dye for fluid leak detection.
- **Actuator Problems**
  - ***Fault Symptoms*** – Unresponsive movement, slow operation, erratic behavior.
  - ***Diagnostic Steps*** – Inspect hydraulic or pneumatic cylinders and motors for internal leaks, rod misalignment, or damaged seals.
  - ***Tools*** – Pressure gauges and manual tests can help verify the actuator's response and function.
- **Contamination**
  - ***Fault Symptoms*** – Slow or erratic movement, stuck components, unusual noises.

- *Diagnostic Steps* – Analyse the fluid for contaminants like metal particles, dirt, or water. Clean or replace filters as necessary.
- *Tools* – Fluid analysis kits, filter testers, and particle counters.

- **Pressure and Flow Issues**
  - *Fault Symptoms* – Loss of power, overheating, unusual noises.
  - *Diagnostic Steps* – Check the system's pressure settings, inspect pressure relief valves, and ensure proper flow control.
  - *Tools* – Pressure and flow meters, relief valve testers.

15.3.1.3 Repairing Faults in Fluid Power Systems

General Repair Techniques

- **Pump Repairs**
  - *Replacement of Worn Parts* – Replace damaged or worn-out pump components like bearings, seals, or rotors.
  - *Realignment* – Correct misalignment between pump and motor to reduce vibrations and wear.
  - *Air Bleeding* – Remove trapped air from hydraulic systems to prevent cavitation and pressure loss.

- **Valve Repairs**
  - *Cleaning and Flushing* – Remove blockages by cleaning the valves or flushing the system with suitable cleaning fluid.
  - *Replacement* – Replace faulty valves or valve components that cannot be repaired.
  - *Adjustment* – Recalibrate or adjust the valve settings if the flow or pressure is incorrect.

- **Hose and Seal Repairs**
  - *Seal Replacement* – Replace damaged seals or O-rings that cause fluid or air leaks.
  - *Hose Tightening* – Tighten loose fittings or connections to stop fluid leaks.
  - *Hose Replacement* – Replace damaged or worn hoses to restore proper pressure and prevent contamination.

- **Actuator Repairs**
  - ***Seal Repair*** – Replace seals or packings in hydraulic or pneumatic cylinders to restore proper movement.
  - ***Cylinder Rebuilding*** – Rebuild or replace components in worn cylinders.
  - ***Lubrication*** – Apply proper lubrication to moving parts in actuators to reduce wear and friction.

- **Contamination Management**
  - ***Filter Replacement*** – Replace clogged filters to ensure the system stays clean.
  - ***System Flush*** – Flush the system to remove contaminants and replace the fluid if necessary.
  - ***Fluid Monitoring*** – Implement regular fluid analysis and monitoring to detect early signs of contamination.

15.3.2 Examples of Fault Analysis and Repairs

15.3.2.1 Hydraulic System (Constant Speed)

- ***Symptom*** – A hydraulic press operates slower than usual and exhibits unusual noises.
- ***Diagnosis*** – Inspection reveals low fluid pressure and air bubbles in the hydraulic lines.
- ***Fault Location*** – Worn-out pump seals and air leaks in the suction line.
- ***Repair*** – Replace pump seals, tighten connections in the suction line, and bleed the air from the system. Conduct a pressure test to ensure normal operation.

15.3.2.2 Pneumatic System (Variable Speed)

- ***Symptom*** – A pneumatic motor exhibits fluctuating speeds and inconsistent torque.
- ***Diagnosis*** – An inspection shows that the control valve is sticking and there is an air leak around the valve.
- ***Fault Location*** – Contaminated control valve and a worn-out O-ring.
- ***Repair*** – Clean the control valve to remove debris, replace the faulty O-ring, and adjust the control system. Perform a flow test to verify normal operation.

15.3.2.3 Hydraulic Lifting System (Variable Speed)

- ***Symptom*** – Hydraulic cylinder in a variable-speed crane is jerky and fails to lift loads smoothly.

- ***Diagnosis*** – The fluid level is low, and there are visible leaks around the cylinder.
- ***Fault Location*** – Internal seal damage in the hydraulic cylinder.
- ***Repair*** – Replace damaged seals, top up the fluid to the recommended level, and conduct a load test to ensure smooth operation.

15.3.3 Preventive Measures to Avoid Faults

1 **Regular Inspection** – Schedule routine checks for leaks, wear, and alignment to catch potential issues early.

2 **Proper Lubrication** – Ensure the right type and amount of lubrication is applied to reduce wear.

3 **Fluid Quality** – Maintain fluid cleanliness with proper filtration and monitoring, and replace fluids periodically.

4 **Proper Training** – Train operators and maintenance personnel on identifying early symptoms and performing minor repairs.

5 **System Monitoring** – Use diagnostic tools like pressure gauges, flow meters, and temperature sensors for early detection of issues.

15.3.4 Summary

In summary, effective troubleshooting and repair in fluid power transmission systems involve understanding the specific characteristics of constant and variable speed devices. Identifying symptoms, locating faults, and implementing the right repairs require a systematic approach, proper tools, and regular maintenance. By following best practices and implementing preventive measures, operators can ensure the longevity, efficiency, and safety of hydraulic and pneumatic systems in industrial applications.

15.4 STATUTORY SAFETY REQUIREMENTS FOR CRANES AND LIFTING TACKLE

Ensuring safety in operations involving cranes and lifting tackle is critical to avoiding accidents, injuries, and equipment damage. Various laws and standards worldwide outline the statutory safety requirements to maintain safety, protect workers, and minimize risks during lifting operations. Below is a summary of the general statutory safety requirements, followed by specific examples based on common standards.

15.4.1 General Statutory Safety Requirements

1 Competent Personnel

- **Operator Certification** – Operators must be certified and adequately trained to use cranes and lifting tackle.
- **Qualified Personnel** – Only competent personnel should perform inspections, maintenance, and repairs on cranes and lifting equipment.

2 Regular Inspection and Testing

- **Pre-Operation Inspection** – Before using cranes and lifting tackle, daily pre-operation checks are mandatory to ensure that the equipment is in safe working condition.
- **Periodic Inspections** – Conduct regular inspections of cranes and lifting gear (monthly, quarterly, or annually) as per regulations. These inspections should be performed by qualified individuals.
- **Load Testing** – Regular load testing is mandatory to verify the crane's capacity to handle specified loads without failure. This is typically done annually or as per local requirements.

3 Safe Working Load (SWL) and Capacity

- **Labeling and Marking** – Cranes and lifting tackle must have visible markings showing their Safe Working Load (SWL) or maximum capacity. Lifting should never exceed the SWL.
- **Load Indicating Devices** – Cranes should have load indicators, limiting devices, or overload alarms to ensure loads do not exceed capacity.

4 Maintenance and Repair

- **Scheduled Maintenance** – Regular and preventative maintenance must be conducted based on the manufacturer's recommendations.
- **Defective Equipment** – Immediately remove any damaged or defective equipment from service until it is repaired or replaced.
- **Maintenance Records** – Maintain comprehensive records of maintenance, inspections, repairs, and load tests.

5 Safety Controls and Devices

- **Emergency Stop** – Cranes must be equipped with emergency stop devices to allow operators to stop operations in case of emergencies.

- **Limit Switches** – Cranes should have limit switches to prevent over-hoisting, over-lowering, and unsafe travel.
- **Safety Brakes** – Use safety brakes to secure loads in case of power failure or mechanical failure.

6 Lifting Tackle Standards

- **Selection of Lifting Tackle** – Choose appropriate lifting tackle (slings, hooks, shackles, chains, etc.) based on the load's weight, size, and nature.
- **Inspection of Lifting Tackle**: Conduct visual inspections before every use to check for wear, deformation, or damage.
- **Storage of Lifting Tackle** – Store lifting equipment properly in designated areas to prevent damage, corrosion, or wear.

7 Safety Precautions During Lifting Operations

- **Safe Loading and Lifting** – Ensure proper rigging, balancing, and securing of the load before lifting.
- **Signalling** – Use standard hand signals, radios, or other communication methods to coordinate lifting operations. Signalers must be qualified and wear high-visibility clothing.
- **Clear Area** – Ensure the load path is clear, and never allow personnel to stand under a suspended load.
- **Load Stability** – Ensure that loads are stable and balanced during lifting and travel.
- **Wind Speed Monitoring** – Monitor wind speed and weather conditions to ensure safe operation, especially for outdoor crane activities.

15.4.2 Specific Statutory Requirements by Region and Standards

a) OSHA (Occupational Safety and Health Administration - USA)

- **Standard**: OSHA 29 CFR 1926 Subpart CC - Cranes and Derricks in Construction
  - *Annual Inspections:* Cranes must be inspected at least annually by a qualified person.
  - *Operator Certification:* Crane operators must be certified or qualified based on experience, training, and evaluation.
  - *Assembly and Disassembly:* Follow specific safety procedures during crane assembly and disassembly, including ensuring stability.

b) LOLER (Lifting Operations and Lifting Equipment Regulations - UK)

- **Inspection Frequency**: Lifting equipment must be thoroughly examined at least every 6 months if used to lift people and annually if used for other purposes.
- **Load Testing**: Conduct load testing if there are significant changes to lifting equipment that could affect its safety.
- **Reporting**: Keep records of thorough examinations and any repairs or maintenance work conducted.
- **Risk Assessment**: Conduct risk assessments before performing lifting operations to identify potential hazards.

c) ASME (American Society of Mechanical Engineers - USA)

- **Standard**: ASME B30.5 - Mobile and Locomotive Cranes
  - ***Pre-Lift Planning:*** Require pre-lift planning, including load calculations, stability assessment, and site preparation.
  - ***Sling Inspection:*** *Inspect* slings regularly and remove them from service if they show signs of wear, damage, or deformation.
  - ***Lifting Personnel:*** Specific guidelines for lifting personnel using cranes, including the use of safety harnesses and fall protection.

d) ISO (International Organization for Standardization)

- **Standard**: ISO 9927 - Cranes: Inspections
  - ***Routine Inspections:*** Daily or pre-use inspections to detect damage or faults.
  - ***Periodic Inspections:*** Monthly to annual inspections depending on the type and frequency of crane use.
  - ***Documentation*** – Maintain comprehensive inspection records for traceability.

15.4.3 Examples of Statutory Requirements in Practice

a) Offshore Oil and Gas Operations

- Cranes used in offshore oil rigs must comply with strict standards such as API (American Petroleum Institute) and ISO to ensure safe operations in hazardous environments.
- They are subject to more frequent inspections due to harsh operating conditions and the potential for severe consequences in case of failures.

b) Construction Sites

- Cranes on construction sites must follow OSHA guidelines, including site-specific assessments for each lift, operator training, and maintaining detailed inspection records.
- Construction cranes are often equipped with warning devices like horns, lights, and alarms to alert personnel about crane movement.

c) Manufacturing Plants

- Overhead cranes used in manufacturing plants are typically inspected according to local regulations, such as LOLER in the UK, with six-monthly checks for lifting personnel and annual checks for lifting materials.
- Maintenance schedules are rigorously followed to ensure that the cranes can handle repetitive use without failures.

15.4.4 Summary

In summary, statutory safety requirements for cranes and lifting tackle are designed to protect workers, equipment, and materials. Compliance with these safety standards involves the correct selection, inspection, maintenance, and usage of cranes and lifting gear. Failing to follow these regulations can lead to severe accidents, loss of life, and significant financial consequences. Each region or industry might have specific regulations, but the core principles of ensuring competent personnel, thorough inspections, adherence to load limits, and regular maintenance are universal in keeping lifting operations safe and efficient.

15.5 MAIN CHARACTERISTICS OF STATIC AND MOBILE TOOLS

In industrial and construction settings, tools are often categorized as **static** or **mobile**. Understanding their characteristics is essential for selecting the right equipment for a particular task. Each type has distinct features, advantages, and limitations based on its application, size, portability, and power source.

15.5.1 Static Tools

Static tools are stationary and are usually fixed in one location. These tools are often large and designed for heavy-duty tasks that require stability, precision, and high power. They are commonly used in manufacturing, workshops, and industrial environments where the workpiece is brought to the tool.

1 Characteristics of Static Tools

- **Stability** – Static tools are anchored or mounted, providing a stable and secure working environment. This allows for precise and consistent operations.
- **Power** – They are typically powered by electricity or hydraulics, delivering high power and efficiency for large-scale or continuous operations.
- **Size and Weight** – Static tools are usually heavy and bulky, which makes them unsuitable for portable applications. They are designed to remain in one place.
- **Precision** – Due to their stable nature, static tools are ideal for detailed, accurate, and repetitive tasks.
- **Capacity** – These tools often have a larger capacity, allowing them to handle bigger or more complex materials.

2 Examples of Static Tools

- **Drill Press** – A drill press is a heavy-duty machine used for precise drilling operations in metalworking or woodworking. Its stability ensures that holes are drilled accurately.
- **Lathe Machine** – A lathe is used for cutting, sanding, drilling, or turning materials like metal or wood. The tool is fixed, while the material rotates, allowing for precise shaping.
- **Bench Grinder** – A bench grinder is mounted on a workbench and is used for grinding, sharpening, or polishing tools and materials. Its fixed position makes it ideal for steady, repetitive work.

15.5.2 Mobile Tools

Mobile tools, also known as portable tools, are designed to be easily moved from one place to another. These tools are lightweight, compact, and usually handheld, allowing flexibility in various work environments. They are ideal for tasks that require mobility or when the tool needs to be brought to the workpiece.

1 Characteristics of Mobile Tools

- **Portability** – Mobile tools are lightweight and easy to carry, making them suitable for on-site jobs and tasks that require movement.
- **Power Sources** – They can be powered by batteries, electricity, pneumatic systems, or fuel, offering versatility depending on the environment.
- **Flexibility** – These tools are adaptable to a wide range of tasks and environments, especially in hard-to-reach areas.

- **Ease of Use** – Mobile tools are designed for quick and efficient use, often featuring ergonomic designs for better handling and comfort.
- **Lower Capacity** – Compared to static tools, mobile tools typically have a lower capacity and are best suited for lighter or more straightforward tasks.

2 Examples of Mobile Tools

- **Cordless Drill** – A cordless drill is a battery-operated, handheld tool used for drilling and fastening screws. Its portability makes it ideal for construction and DIY projects.
- **Angle Grinder** – An angle grinder is a handheld power tool used for cutting, grinding, and polishing materials. Its portability allows it to be used on construction sites or in confined spaces.
- **Impact Wrench** – An impact wrench is a portable tool that uses compressed air or batteries to deliver high torque for loosening or tightening bolts. It is commonly used in automotive and construction applications.

15.5.2.1 Table of Comparison Between Static and Mobile Tools

Characteristic	**Static Tools**	**Mobile Tools**
Mobility	Fixed in one location	Portable and easy to move
Power Source	Typically, electrical or hydraulic	Battery, electric, pneumatic, or fuel-powered
Stability	High stability for precision work	Less stable, designed for flexibility
Capacity	Larger capacity, suitable for heavy-duty tasks	Smaller capacity, best for light to medium tasks
Usage	Repetitive, detailed, and precise operations	Quick, versatile, and adaptable use
Examples	Drill Press, Lathe, Bench Grinder	Cordless Drill, Angle Grinder, Impact Wrench

15.5.3 Summary

In summary, the choice between static and mobile tools depends on the nature of the job, workspace, and specific requirements. Static tools are preferred for accuracy, power, and large-scale operations, while mobile tools are ideal for tasks requiring flexibility, portability, and accessibility. Proper selection and usage of these tools can enhance efficiency, safety, and productivity in various industries, from manufacturing to construction and maintenance.

CHAPTER SIXTEEN

16.0 COMPREHENSIVE GUIDE TO PUMP AND STEAM TRAP INSTALLATION, MAINTENANCE, AND PIPE CIRCUIT FAULT DIAGNOSIS

16.1 PROCEDURE FOR THE INSTALLATION OF PUMPS

The installation of pumps involves several crucial steps to ensure proper functioning, efficiency, and longevity.

16.1.1 Step-by-Step Guide for Pump Installation

1 Pre-Installation Preparation

Before starting the installation, it's essential to conduct a thorough assessment of the site and equipment to ensure that all requirements are met:

- **Site Inspection** – Check the installation site for any obstacles or hazards. Ensure the surface is level and free of debris.
- **Pump and Equipment Check** – Inspect the pump for any damage or defects. Ensure that all necessary components and accessories (such as gaskets, bolts, and couplings) are available.
- **Foundation Preparation** – If the pump requires a foundation, such as a concrete base, ensure that it is level, sturdy, and can handle the pump's weight and vibrations.
- **Required Tools** – Gather the necessary tools and equipment for the installation, including wrenches, alignment tools, sealant, level, and measuring tape.

Example – Installing a centrifugal pump in an industrial setting requires a reinforced concrete base to support the weight and handle vibrations during operation.

2 Positioning the Pump

Proper positioning is crucial to avoid future problems related to alignment and performance:

- **Position the Pump** – Place the pump on its designated base or foundation. Ensure it is centered and level. Use lifting equipment like hoists or cranes for heavy pumps.
- **Leveling** – Check the pump's level with a spirit level and make adjustments using shims if necessary. A level installation prevents strain on the motor and reduces wear.

Example – For a submersible pump, ensure the pump is submerged to the correct depth in the liquid, ensuring that it's securely fastened to prevent movement during operation.

3 Aligning the Pump and Motor

Proper alignment between the pump and the motor is critical to avoid operational issues like vibrations or premature wear:

- **Coupling Alignment** – Align the pump and motor shafts using alignment tools such as a dial indicator or laser alignment device. Misalignment can cause excessive wear and lead to operational failures.
- **Angular and Parallel Alignment** – Ensure both angular and parallel alignment by checking the gap and positioning between the couplings. Adjust as necessary.
- **Securing Bolts** – Tighten the foundation bolts gradually while checking alignment to ensure it doesn't shift during the tightening process.

Example – For an inline booster pump, proper alignment is critical to ensure efficient pressure boosting and prevent leaks at the connection points.

4 Piping Installation

Proper piping setup is essential to ensure smooth fluid flow and reduce mechanical stress on the pump:

- **Connect Inlet and Outlet Pipes** – Attach the inlet and outlet pipes, ensuring that they are properly aligned with the pump. Avoid using pipes that exert strain on the pump.
- **Support Piping** – Use pipe supports to prevent the weight of the pipes from resting on the pump. Flexible connectors or expansion joints can help accommodate thermal expansion or vibration.
- **Check for Leakage** – Ensure all connections are tight and secure. Use gaskets, thread sealant, or tape to prevent leaks at joints.

Example – Installing a diaphragm pump requires ensuring that the suction and discharge lines are appropriately connected, with minimal bends to reduce pressure loss.

5 Electrical Connections

Proper electrical setup is necessary to ensure the pump operates efficiently and safely:

- **Power Supply** – Verify that the power supply matches the pump's voltage and phase requirements. Check that the main power is off before making any connections.

- **Wiring** – Connect the pump's electrical wiring according to the manufacturer's instructions. Ensure that the wiring is protected, and use appropriate conduits if necessary.
- **Grounding** – Properly ground the pump to prevent electrical hazards.

Example – When installing an electric centrifugal pump, ensure that the wiring is correctly done to avoid motor burnout or electrical failures.

6 Priming and Lubrication

Priming and lubrication help prepare the pump for startup and ensure smooth operation:

- **Priming the Pump** – If the pump is not self-priming, fill it with liquid to eliminate air pockets. This is especially important for centrifugal pumps, which can be damaged if run dry.
- **Lubricating Bearings** – Check and fill the bearings with the recommended lubricant. Proper lubrication reduces friction and prolongs the life of the pump.

Example – In gear pumps, the initial priming with the operating fluid ensures that the gears are lubricated before starting.

7 Testing and Start-Up

The final steps involve testing the pump to ensure everything is functioning correctly:

- **Rotation Check** – Before starting, check the pump's rotation to ensure that the motor is driving the pump in the correct direction. Incorrect rotation can damage some pumps.
- **Start-Up** – Gradually start the pump, monitoring for any unusual noises, vibrations, or leaks. If any issues are observed, shut down and adjust accordingly.
- **Performance Testing** – Measure the pump's performance parameters, such as flow rate, pressure, and efficiency, to ensure they match the manufacturer's specifications.

Example – For a fire-fighting pump, performance testing involves checking the pressure output and ensuring that it meets the required standards for fire suppression.

8 Final Inspection and Documentation

Once the pump is installed and tested:

- **Inspection** – Conduct a final inspection to verify that everything is in order, including alignment, piping, and electrical connections.
- **Documentation** – Record the installation details, including any adjustments made, tests performed, and maintenance recommendations. Keep a record of any warranty or manufacturer documentation.

Example – For an HVAC circulation pump, the final inspection includes checking flow and pressure, documenting results, and ensuring proper flow to all parts of the HVAC system.

16.1.2 Summary

In summary, proper pump installation is vital for ensuring the equipment's longevity, efficiency, and safety. Each step, from preparation to final inspection, should be meticulously followed to prevent operational problems and enhance reliability.

16.2 FOUNDATION ALIGNMENT PROCESS

Foundation alignment is crucial for the stability and operational efficiency of machinery and equipment. Proper alignment ensures that equipment is correctly positioned, reducing vibration, wear, and tear, and ensuring safe operation. Below, I'll demonstrate the step-by-step process of foundation alignment, along with specific examples for better understanding.

16.2.1 Step-By-Step Process of Foundation Alignment

Step 1 – Preparing the Foundation

Before starting the alignment, the foundation must be adequately prepared:

- **Site Cleaning** – Clean the foundation surface, removing any debris, oil, or dirt that could affect the equipment's placement.
- **Check Level** – Use a spirit level to check the flatness of the foundation. If the surface is uneven, it may require additional leveling, such as using a grinder or adding leveling grout.
- **Install Anchor Bolts** – Ensure that the anchor bolts are in the correct position according to the manufacturer's specifications. These bolts will hold the equipment firmly in place.

Example – In the installation of a large electric motor, the concrete base should be free of cracks, dirt, or oil stains that could affect alignment.

Step 2: Setting Up the Equipment on the Foundation

Once the foundation is clean and level, the next step is to position the equipment:

- **Placement of Equipment** – Use a crane, hoist, or other lifting equipment to place the machine or equipment onto the foundation. Ensure that it is in the correct orientation based on layout specifications.
- **Initial Alignment** – Perform a rough alignment using the layout drawings, ensuring the equipment is centered and positioned according to the foundation markings.

Example – Positioning a centrifugal pump on a concrete base requires the use of a crane to lift the pump and place it on pre-installed anchor bolts.

Step 3: Checking and Adjusting the Horizontal Alignment

The horizontal alignment ensures that the equipment is level and in line with the foundation's horizontal plane.

- **Use of Spirit Level** – Place a spirit level or precision level on the equipment's base to verify that it is horizontally level. The bubble should be centered in the level's vial.
- **Shim Adjustment** – Use shims (thin metal or plastic plates) to make adjustments. Insert shims under the base of the equipment as needed until it is perfectly level.
- **Tighten Bolts** – Gradually tighten the anchor bolts while checking the alignment. Tighten each bolt evenly to avoid shifting the equipment.

Example – For an HVAC unit on a steel platform, shims can be inserted under the mounting brackets to ensure that the unit is level horizontally.

Step 4: Checking Vertical Alignment

Vertical alignment is crucial to ensure that the equipment is not tilted or leaning:

- **Plumb Line or Laser Level** – Use a plumb line or laser level to check that the equipment is vertically aligned. The tool should indicate that the equipment is perfectly upright.
- **Vertical Adjustments** – If necessary, adjust the vertical position by changing the thickness of the shims under specific areas of the equipment's base. Ensure even distribution to avoid tilting.

Example – In the installation of a compressor, using a laser level will help verify that the machine is perpendicular to the floor, avoiding tilting that could cause operational issues.

Step 5: Precision Alignment with Dial Indicator or Laser Alignment Tools

For more critical or precise installations, such as rotating machinery, a more accurate alignment tool is necessary:

- **Dial Indicator Setup** – Attach a dial indicator to the equipment to measure the misalignment. Rotate the indicator around the equipment's axis, recording any deviations.
- **Laser Alignment** – Use laser alignment tools to check the alignment between the equipment and its couplings. Adjust as needed to ensure that the laser readings are within the manufacturer's specified tolerance.

- **Final Adjustments** – Adjust the position of the equipment based on the measurements taken until the alignment is within acceptable tolerances.

Example – Aligning a motor and pump coupling requires the use of a dial indicator to measure any misalignment between the two components. Correct alignment reduces wear on bearings and shafts.

Step 6: Securing and Re-checking Alignment

Once the alignment is complete, securing the equipment is essential:

- **Tighten Anchor Bolts** – Securely tighten the anchor bolts in a cross-pattern sequence to maintain even pressure. Use a torque wrench to ensure bolts are tightened to the correct specifications.
- **Re-check Alignment** – After tightening the bolts, re-check the horizontal and vertical alignment using the same tools to ensure that the equipment has not shifted during the tightening process.
- **Fill Grout** – For heavy equipment, fill any gaps between the equipment base and foundation with non-shrink grout. Allow the grout to cure before the final check.

Example – When installing a large gear reducer, it's critical to re-check alignment after tightening bolts, as the alignment can shift slightly during the bolting process.

Step 7: Final Alignment Check After Load Application

After the equipment has been placed under operational load, perform a final alignment check:

- **Operational Load Test** – Operate the equipment under normal conditions to check for any unusual vibrations or noises. This test ensures the foundation and alignment can handle operational stresses.
- **Monitor Alignment** – Use monitoring tools, such as vibration sensors, to detect any alignment deviations during operation.
- **Fine Adjustments** – Make any fine adjustments as needed after the load test to ensure the alignment remains within the acceptable range.

Example – For an industrial fan installation, check the vibration levels during operation, as misalignment can cause increased vibration, leading to wear and failure.

16.2.2 Summary

In summary, proper foundation, alignment is critical for ensuring the stability, efficiency, and longevity of machinery and equipment. Each step, from preparation to final checks, must be executed carefully to prevent operational issues like vibration, premature wear, and safety risks. Properly aligned equipment requires less maintenance, runs more efficiently, and has a longer lifespan.

16.3 MAINTENANCE SCHEDULE FOR PUMPS

Creating a maintenance schedule for pumps is essential to ensure their efficiency, longevity, and reliable performance. A well-structured maintenance schedule helps to prevent unexpected breakdowns, reduce operational costs, and extend the life of the equipment.

16.3.1 Breakdown of a Typical Maintenance Schedule for Pumps

Daily Maintenance Tasks

1 Visual Inspection

- Inspect the pump for any visible leaks, unusual noises, or vibrations.
- Check the seals, gaskets, and connections for any signs of wear or leakage.
- Ensure that the pump is clean and free from any debris that could affect its operation.

2 Check Operational Parameters

- Monitor and record the flow rate, pressure, and other key operational parameters.
- Verify that the pump is operating within the recommended temperature range.
- Ensure that the power consumption is normal, indicating the pump is not under strain.

3 Lubrication

- Check the lubrication levels (where applicable) and add oil or grease if needed.

Weekly Maintenance Tasks

1 Check for Unusual Vibrations

- Use a vibration analyser or manual check to detect any abnormal vibrations.
- Investigate any unusual sounds or vibrations to determine if they are caused by misalignment or worn components.

2 Inspect and Clean Filters/Strainers

- Check the condition of filters and strainers to ensure there are no blockages.
- Clean or replace filters if clogged or damaged.

3 **Check Belt and Coupling Tension**

- Inspect the belt (if applicable) and check the tension. Adjust if necessary.
- Examine couplings for wear and proper alignment.

Monthly Maintenance Tasks

1 **Alignment Check**

- Perform a detailed alignment check between the pump and motor using alignment tools (dial indicator or laser alignment tools).
- Adjust alignment if necessary to prevent premature wear and reduce vibrations.

2 **Test Pump Protection Devices**

- Check the functioning of pump protection devices such as pressure gauges, temperature sensors, and vibration sensors.
- Calibrate devices if necessary to ensure accurate readings.

3 **Inspect Seals and Packing**

- Check the mechanical seals or packing for leaks or wear.
- Tighten or replace seals as needed to prevent leakage.

Quarterly Maintenance Tasks

1 **Detailed Performance Assessment**

- Conduct a performance test to compare the pump's current efficiency with its baseline or manufacturer's specifications.
- Record flow rate, pressure, power consumption, and temperature for trend analysis.

2 **Drain and Inspect Lubrication Oil (if applicable)**

- Drain and inspect the oil for any contamination or degradation.
- Refill with fresh oil according to manufacturer specifications.
- Check lubrication points for proper functioning.

3 **Inspect Electrical Connections**

- Inspect the electrical wiring, terminals, and connections for corrosion or loose fittings.
- Tighten connections and ensure that the motor insulation is in good condition.

4 **Pump Bearings Check**

- Examine pump bearings for wear or damage.
- Re-lubricate or replace bearings if they show signs of wear.

Annual Maintenance Tasks

1 **Complete Disassembly and Inspection**

- Conduct a complete disassembly of the pump for a thorough inspection.
- Inspect all internal components such as impellers, casings, shafts, and seals for wear, corrosion, or damage.
- Replace any worn-out parts with genuine spare parts as needed.

2 **Clean and Inspect the Pump Casing**

- Remove any buildup or scaling inside the pump casing.
- Inspect the casing for cracks or corrosion. If damaged, replace or repair the casing.

3 **Check Motor and Electrical Systems**

- Conduct a comprehensive motor inspection, including the insulation resistance test.
- Clean the motor windings and check the motor bearings.
- Inspect the control panel, switches, and overload protection devices for any faults.

4 **Hydrostatic Testing**

- Perform a hydrostatic pressure test to ensure that the pump casing can handle the operating pressure without leaks.

5 **Calibration and Re-commissioning**

- Calibrate all monitoring instruments such as flow meters, pressure gauges, and temperature sensors.
- Re-commission the pump after completing maintenance tasks to ensure it is operating efficiently.

16.3.1.1 Table of Maintenance Schedule Overview

Frequency	Task	Description
Daily	Visual inspection, operational parameter check, lubrication	Routine checks for leaks, noises, and lubrication levels
Weekly	Vibration check, filter cleaning, belt and coupling inspection	Checking for vibrations, adjusting belts, cleaning filters
Monthly	Alignment check, seal inspection, protection devices test	Ensure proper alignment and inspect seals, test protection devices
Quarterly	Performance assessment, lubrication inspection, electrical connections, bearing check	In-depth performance checks, lubrication inspection, and checking electrical connections
Annual	Complete disassembly, inspection, hydrostatic test, motor inspection, calibration	Comprehensive overhaul, pressure testing, motor check, and calibration

16.3.2 Examples of Pump Maintenance

1 Centrifugal Pump Maintenance

- **Daily** – Check for unusual noises or vibrations.
- **Monthly** – Test the impeller for wear and inspect the casing.
- **Annual** – Conduct a full disassembly, inspect impeller wear, and replace gaskets if necessary.

2 Submersible Pump Maintenance

- **Weekly** – Check the intake screens for blockages.
- **Quarterly** – Inspect the electrical wiring and waterproof seals for any damage.
- **Annual** – Remove and inspect the entire pump to check for wear in the motor and impeller.

3 Gear Pump Maintenance

- **Daily** – Ensure the gears are properly lubricated.
- **Monthly** – Inspect the shaft and bearings for wear.
- **Annual** – Disassemble the gear casing to check for wear on gears and clean any buildup.

16.3.3 Summary

In summary, regular maintenance of pumps, following a structured schedule, helps ensure the equipment's reliability, safety, and longevity. This proactive approach minimizes the risk of unexpected failures and extends the pump's service life, providing better efficiency and lower operational costs.

16.4 DIAGNOSING AND REPAIRING FAULTS IN PUMPS

Diagnosing and repairing faults in pumps is essential for maintaining operational efficiency and avoiding costly downtime.

16.4.1 Pump Troubleshooting: Symptoms, Causes, Diagnosis, and Repair Solutions

1 Symptom: Reduced or No Flow

When a pump exhibits reduced flow or completely stops delivering fluid, several factors may be responsible.

Potential Causes

- **Clogged Filters or Strainers** – Dirt, debris, or foreign particles may have blocked the intake or discharge.
- **Air Leaks** – Suction line air leaks can cause air to enter the system, reducing efficiency.
- **Worn Impeller** – The impeller may be damaged, corroded, or worn out, reducing pumping efficiency.
- **Cavitations** – Cavitation occurs when vapor bubbles form in the pump due to low pressure or improper suction conditions.

Fault Identification

- **Check Filters** – Inspect the intake filters or strainers for blockages.
- **Pressure Test** – Perform a pressure test on the suction line to identify leaks.
- **Visual Inspection** – Open the pump casing and visually inspect the impeller for signs of damage or corrosion.
- **Vibration Analysis** – Use a vibration meter to detect cavitation (characterized by high-frequency vibrations).

Repair Solutions

- **Clean or Replace Filters** – Remove any debris from filters or replace clogged strainers.

- **Seal Air Leaks** – Repair or replace leaking gaskets, seals, or joints in the suction line.
- **Impeller Replacement** – Replace worn or damaged impellers with new ones.
- **Adjust Suction Conditions** – Increase suction pressure or reduce pump speed to minimize cavitation.

2 Symptom: Excessive Vibration or Noise

Unusual vibrations or noises from the pump indicate potential mechanical or operational faults.

a) Potential Causes

- **Misalignment** – The pump and motor are not properly aligned, causing excessive vibration.
- **Loose Components** – Bolts, bearings, or internal parts may have loosened over time.
- **Bearing Failure** – Worn or damaged bearings can cause high noise levels.
- **Cavitation** – Air bubbles collapsing within the pump create noise.

b) Fault Identification

- **Alignment Check** – Use a laser alignment tool or dial indicator to check if the pump and motor shafts are correctly aligned.
- **Component Inspection** – Visually inspect all bolts, bearings, and internal components for tightness.
- **Bearing Analysis** – Use a stethoscope or vibration analyser to detect bearing noise and identify wear.

c) Repair Solutions

- **Re-align Pump and Motor** – Use an alignment tool to correct any misalignment.
- **Tighten or Replace Components** – Tighten loose bolts or replace worn internal parts.
- **Replace Bearings** – Remove and replace damaged or worn bearings with appropriate replacements.
- **Adjust Suction Conditions** – Reduce pump speed or increase suction pressure if cavitation is detected.

3 Symptom: Overheating

If a pump is overheating, it may indicate a mechanical, electrical, or operational issue.

a) **Potential Causes**

- **Blocked Suction or Discharge** – Obstructions in the pump lines can force the motor to work harder, leading to overheating.
- **Worn Bearings** – Excess friction due to worn bearings can cause heat buildup.
- **Incorrect Lubrication** – Insufficient or improper lubrication of moving parts increases friction.
- **Electrical Issues** – Overloading or electrical faults in the motor can cause overheating.

b) **Fault Identification**

- **Flow Inspection** – Check the suction and discharge lines for any obstructions or partial blockages.
- **Bearing Check** – Inspect the bearings for wear and test lubrication levels.
- **Electrical Diagnostics** – Use a multimeter to check the electrical supply, wiring, and motor windings.

c) **Repair Solutions**

- **Clear Blockages** – Remove any obstructions in the suction or discharge lines.
- **Replace Bearings** – Install new bearings if the existing ones are damaged or worn.
- **Lubricate Properly** – Add the correct type and quantity of lubrication according to the manufacturer's specifications.
- **Check Electrical Components** – Replace faulty electrical components, cables, or wires.

4 Symptom: Pump is Not Priming

Priming issues can prevent the pump from creating the necessary vacuum for fluid movement.

a) **Potential Causes**

- **Suction Line Leaks** – Air leaks in the suction line can prevent proper priming.
- **Blocked Suction** – Clogged suction filters or intake lines can obstruct flow.
- **Improper Installation** – Incorrect installation of the pump can create priming problems.
- **Worn Seals** – Damaged seals can cause air to leak into the system, reducing priming efficiency.

b) **Fault Identification**

- **Air Leak Test** – Conduct an air leak test on the suction line using a pressure gauge.

- **Suction Inspection** – Inspect the suction line and filters for blockages.
- **Installation Review** – Verify that the pump's installation follows manufacturer guidelines.
- **Seal Check** – Inspect the pump's seals and gaskets for wear or damage.

Repair Solutions

- **Seal Leaks** – Repair or replace damaged seals and gaskets.
- **Clear Blockages** – Remove any obstructions in the suction line or filters.
- **Re-install Pump** – Adjust the installation to meet manufacturer specifications if installation errors are detected.

5 Symptom: Reduced Pressure or No Pressure

A reduction in pump pressure or complete loss of pressure can indicate several issues.

Potential Causes

- **Impeller Damage** – A damaged or worn impeller can reduce the pump's pressure generation.
- **Air in System** – Air entrained in the pump reduces efficiency and pressure.
- **Wear in Pump Components** – Components like diffusers or volutes may be worn.
- **Suction Lift Issues** – High suction lift can cause low-pressure output.

Fault Identification

- **Impeller Inspection** – Open the pump casing and inspect the impeller for wear or damage.
- **Air Check** – Look for signs of air bubbles in the system or check if the suction line is drawing in air.
- **Component Inspection** – Inspect all internal components for signs of wear.
- **Suction Test** – Measure suction lift to ensure it is within the specified range.

Repair Solutions

- **Replace Impeller** – Install a new impeller if the existing one is damaged or worn.
- **Purge Air** – Purge air from the system and repair any leaks in the suction line.
- **Component Replacement**: Replace worn-out components like diffusers or volutes.
- **Adjust Suction Lift** – Modify suction lift to stay within acceptable limits.

6 Symptom: Pump Does Not Start

If the pump does not start, the issue is likely electrical or mechanical.

Potential Causes

- **Power Supply Issues** – A faulty power supply or electrical connections can prevent the pump from starting.
- **Motor Failure** – A failed motor or burnt motor windings can cause a failure to start.
- **Tripped Circuit Breaker** – Overloading or a fault in the circuit may trip the breaker.
- **Jammed Impeller** – A jammed impeller can prevent the motor from turning over.

Fault Identification

- **Electrical Check** – Use a multimeter to check the power supply, fuses, and connections.
- **Motor Diagnostics** – Inspect the motor for signs of burnout or mechanical failure.
- **Circuit Inspection** – Inspect the circuit breaker for any faults or overloads.
- **Impeller Inspection** – Check if the impeller is blocked or jammed with debris.

Repair Solutions

- **Repair Electrical Connections** – Fix faulty wiring, replace blown fuses, or correct electrical faults.
- **Motor Replacement** – Replace the motor if it is burnt out or not functioning.
- **Reset Circuit Breaker** – Reset the breaker if tripped and investigate the cause of overload.
- **Clear Jam** – Remove any debris blocking the impeller and clean the pump.

16.4.2 Summary

In summary, properly identifying and addressing pump faults requires a methodical approach. A combination of visual inspections, diagnostic tools, and regular maintenance helps to ensure the longevity and efficiency of pumps. A well-maintained pump reduces the likelihood of operational disruptions and minimizes repair costs over time.

16.5 PROCEDURES FOR INSTALLATION AND MAINTENANCE OF STEAM TRAPS

The proper installation and maintenance of steam traps are crucial for ensuring the efficiency of steam systems, minimizing energy loss, and protecting equipment. Steam traps are used to discharge condensate, air, and non-condensable gases from steam systems without releasing steam.

16.5.1 Step – By – Step Installation Procedure for Steam Traps.

Proper installation is key to the efficient operation of steam traps. Following a systematic approach ensures that the steam trap functions effectively, maintaining optimal steam system performance.

1 Select the Right Steam Trap

- Choose the appropriate type of steam trap for the application (e.g., thermodynamic, thermostatic, float and thermostatic, or inverted bucket).
- Consider factors like pressure range, temperature, load capacity, and environmental conditions.

2 Locate the Installation Point

- Install the steam trap at the lowest point of the equipment or piping to allow gravity drainage of condensate.
- Ensure the steam trap is located in an accessible area for inspection and maintenance.
- Avoid installing the steam trap at points where dirt or debris can accumulate, unless a strainer is used.

3 Install Isolation Valves

- Place isolation valves upstream and downstream of the steam trap to allow for safe maintenance without shutting down the entire system.
- Use gate or globe valves for isolation, and ensure they are of the same pressure rating as the steam system.

4 Install a Strainer (Optional but Recommended)

- Use a strainer upstream of the steam trap to filter out dirt and debris.
- Ensure the strainer has a blowdown valve to facilitate easy cleaning during maintenance.

5 Install the Steam Trap Horizontally (If Applicable)

- Ensure that the steam trap is installed horizontally, unless it is specifically designed for vertical installation.
- Confirm that the flow direction matches the arrow indicated on the steam trap body.

6 Ensure Proper Piping Arrangement

- Maintain a slight slope in the condensate piping toward the steam trap to facilitate drainage.
- Use short, direct piping runs to minimize pressure drop and maximize efficiency.
- Use reducers to ensure the proper pipe size if necessary.

7 Check the Venting

- Provide adequate venting for air and non-condensable gases, especially when using thermostatic steam traps.
- Install a vacuum breaker, if necessary, to prevent the formation of a vacuum in the system.

8 Install Test Valve or Test Port (Optional)

- Include a test valve or test port downstream of the steam trap to verify its operation and discharge.

9 Leak Test the Installation

- Pressurize the system and check for any leaks around the steam trap connections, joints, and valves.
- Tighten connections if necessary, and ensure no steam is escaping.

16.5.2 Step-by-Step Maintenance Procedure for Steam Traps

Regular maintenance is essential to keep steam traps operating effectively, avoid energy losses, and prevent equipment damage. Maintenance should involve routine inspections, cleaning, and repairs if needed.

1 Visual Inspection

- Perform regular visual inspections of the steam traps to check for leaks, physical damage, or any abnormal discharge.
- Look for signs of wear, corrosion, or steam leaks around the steam trap and connection points.

2 Listen for Unusual Noises

- Use an ultrasonic listening device or stethoscope to detect unusual sounds that may indicate issues.

- A properly functioning steam trap should have a periodic discharge sound. A continuous or intermittent hissing sound may indicate a failed open trap.

3 **Temperature Monitoring**

- Use an infrared temperature gun or thermal imaging camera to measure the inlet and outlet temperatures of the steam trap.
- A significant temperature difference could indicate proper functioning, while no difference may suggest a blockage or failure.

4 **Testing the Steam Trap**

- Use the installed test valve or test port to monitor the discharge pattern of the steam trap.
- Perform a blowdown test to observe the discharge behavior. A healthy steam trap should remove condensate without releasing live steam.

5 **Cleaning the Strainer**

- If a strainer is installed, clean it regularly to remove dirt, debris, and scale.
- Open the strainer's blowdown valve to flush out collected dirt, and clean the strainer mesh with a wire brush if necessary.

6 **Check for Waterlogging**

- Ensure that the steam trap is not waterlogged, which can happen due to a blocked or undersized trap.
- Waterlogging may indicate a failed closed trap or improper sizing, requiring immediate attention.

7 **Repair or Replace Worn Parts**

- Inspect internal components such as discs, seats, float mechanisms, and thermostatic elements for wear.
- Replace worn or damaged parts with genuine replacement parts from the manufacturer.
- Tighten loose connections and replace any leaking gaskets or seals.

8 **Test and Verify Operation**

- After completing maintenance, test the steam trap's operation by re-pressurizing the system.

- Monitor the discharge pattern and temperature to ensure that the steam trap is functioning correctly.

9 Record Maintenance Activities

- Keep a maintenance log of inspections, repairs, and replacements.
- Record the date, type of maintenance performed, and any observations related to the steam trap's performance.
- Use this information to establish maintenance trends and predict future needs.

16.5.2.1 Table of Example of Maintenance Schedule for Steam Traps

Frequency	Task	Description
Monthly	Visual Inspection, Temperature Check	Routine checks for leaks, abnormal noises, and temperature variations.
Quarterly	Strainer Cleaning, Testing	Clean strainer, check discharge behavior using test port, and listen for unusual sounds.
Semi-Annually	Component Inspection, Seal Check	Inspect internal components for wear, replace seals if necessary.
Annually	Complete Disassembly, Overhaul	Conduct full disassembly, inspect all internal parts, clean components, and replace damaged parts.

16.5.3 Common Issues in Steam Traps and Solutions

1 Failed Open Trap

- ***Symptom*** – Continuous steam discharge.
- ***Cause*** – Worn seat, disc, or mechanism.
- ***Solution*** – Replace the faulty part, such as the seat or disc.

2 Failed Closed Trap

- ***Symptom*** – No condensate discharge, waterlogging, or increased backpressure.
- ***Cause*** – Blocked strainer, jammed mechanism, or dirt accumulation.
- ***Solution*** – Clean the strainer, inspect the mechanism, and remove blockages.

3 **Excessive Blow-Through**

- ***Symptom*** – Steam leakage along with condensate discharge.
- ***Cause*** – Damaged internal components, improper sizing.
- ***Solution*** – Replace worn internal parts, ensure the steam trap is correctly sized.

4 **Water Hammer**

- ***Symptom*** – Loud banging noise in the pipes.
- ***Cause*** – Improper installation or undersized trap.
- ***Solution*** – Re-align or re-size the steam trap, adjust piping if necessary.

16.5.4 Summary

In summary, proper installation and regular maintenance of steam traps are essential for efficient steam system operation. Following the outlined procedures helps to minimize energy loss, improve system efficiency, and ensure safe operation. Regular inspection and timely replacement of worn parts help avoid costly repairs and unexpected downtimes, keeping the steam system running optimally.

16.6 SYMPTOMS AND FAULTS IN PIPE CIRCUITS

In a piping system, understanding the relationship between observed symptoms and potential faults is crucial for diagnosing and maintaining efficient operation. Below is a detailed breakdown of common symptoms in pipe circuits and the associated faults that could be causing them.

16.6.1 Common Symptoms and Associated Faults in Pipe Circuits

1 **Symptom – Low or Inconsistent Pressure**

- **Possible Faults**
  - **Leakage in the Pipe Circuit** – Caused by cracks, corrosion, loose joints, or damaged seals, resulting in pressure loss.
  - **Blocked or Partially Obstructed Pipe** – Sediment, scale, or debris buildup inside the pipe can restrict flow and reduce pressure.
  - **Worn or Damaged Pump** – A faulty pump can fail to maintain consistent pressure throughout the system.

- **Valve Malfunction** – Stuck, improperly calibrated, or partially closed valves can restrict flow and reduce pressure.

2 **Symptom: Excessive Noise (e.g., Water Hammer, Whistling)**

- **Possible Faults**
  - **Water Hammer** – A loud banging noise caused by sudden changes in pressure due to rapid valve closures or pump failures.
  - **High Velocity Flow** – If flow speed is too high, it can cause whistling or vibration noises in the pipes.
  - **Cavitation** – Occurs when pressure drops below the fluid's vapor pressure, forming vapor bubbles that collapse, causing noise.
  - **Loose Components** – Vibration due to loose or improperly secured pipes can lead to rattling and noise.

3 **Symptom: Unusual Vibration in the Piping System**

- **Possible Faults**
  - **Misaligned Pipe Supports** – Improperly installed or damaged supports can lead to excessive vibration.
  - **Pump or Equipment Imbalance** – Imbalanced rotating components like pumps or compressors can cause vibration to transfer to the pipes.
  - **Inadequate Damping** – Lack of vibration dampeners or flexible connections can increase system vibrations.
  - **Water Hammer** – A pressure surge can cause vibration in the pipes when valves close rapidly.

4 **Symptom: Fluid Leaks (Visible Dampness or Pooling)**

- **Possible Faults**
  - **Cracked or Corroded Pipes** – Metal fatigue, corrosion, or impact damage can lead to visible leaks.
  - **Loose Fittings or Flanges** – Poorly tightened connections or damaged gaskets can cause fluid leaks.
  - **Damaged Seals** – O-rings, seals, or gaskets that have degraded or are improperly installed can result in leaks.

- **Valve Packing Failure** – If the valve packing has worn out, it can lead to leaks around the valve stem.

5 **Symptom: High Fluid Temperature or Overheating**

- **Possible Faults**
  - **Blocked Heat Exchanger** – A fouled or blocked heat exchanger can prevent proper cooling, leading to elevated temperatures.
  - **Faulty Temperature Control Valve** – A malfunctioning control valve may allow excessive hot fluid into the system.
  - **Insufficient Insulation** – Poor or damaged insulation on hot water pipes can result in heat loss and temperature increase in adjacent pipes.
  - **Pump Failure** – If the pump is not circulating fluid effectively, localized heating can occur in the system.

6 **Symptom: Low Flow Rate or No Flow**

- **Possible Faults**
  - **Obstruction in the Pipe** – Blockages due to debris, scale, or sludge buildup can limit flow.
  - **Closed or Partially Closed Valve** – Valves not fully open can restrict or stop the flow entirely.
  - **Pump Failure** – A malfunctioning pump may not generate enough force to maintain the desired flow rate.
  - **Collapsed Pipe** – A section of the pipe may have collapsed due to external pressure or damage, restricting flow.

7 **Symptom: Sudden Increase in Energy Consumption**

- **Possible Faults**
  - **Pump Running Inefficiently** – A worn or damaged pump may consume more power to maintain the same flow rate.
  - **Leaks in the System** – Undetected leaks can cause pumps to overwork, leading to higher energy consumption.
  - **Incorrect Valve Settings** – If valves are not properly set, it may increase resistance, requiring more power.

- **Blockages or Restrictions** – A partially blocked system forces pumps to work harder to maintain flow, consuming more energy.

8 **Symptom: Erratic or Fluctuating Flow**

- **Possible Faults**
- **Air Traps or Air Bubbles** – Air pockets in the system can cause uneven flow, leading to fluctuations.
- **Faulty or Intermittently Failing Valve** – Valves that are not operating smoothly can cause erratic flow.
- **Inconsistent Pump Performance** – A pump that cycles on and off or has variable speed issues can lead to fluctuating flow.
- **Improper Pipe Sizing** – Pipes that are too small or improperly sized can create turbulent flow conditions.

9 **Symptom: Corrosion or Discoloration in the Pipes**

- **Possible Faults**
- **Electrochemical Corrosion** – Dissimilar metals in contact or stray electrical currents can cause corrosion.
- **Poor Water Quality** – Water with high mineral content, chemicals, or pH imbalance can lead to corrosion.
- **Inadequate Protection** – Lack of proper coatings or inhibitors can accelerate corrosion in metal pipes.
- **Pipe Material Incompatibility** – Using incompatible materials in the system can lead to galvanic corrosion.

16.6.2 Examples of Symptoms and Associated Faults in Pipe Systems

- **Example 1**: A factory notices low pressure in their process water line. After inspection, they discover scale buildup in the main feed line, which was causing restricted flow and a drop in pressure.
- **Example 2**: A manufacturing plant reports unusual vibration near a valve station. Upon diagnosis, technicians find that the valve is closing too quickly, causing water hammer. The solution involves installing a dampening valve and adjusting the closing speed.

- **Example 3**: In an HVAC system, technicians notice excessive noise from the piping network. They locate the problem as air pockets trapped in the system. Bleeding the air from the pipes restores normal operation and eliminates the noise.
- **Example 4**: A pumping station has increased energy consumption without a rise in output. On investigation, a partially blocked filter is found, causing the pump to work harder. Cleaning the filter resolves the issue.

16.6.3 Tips for Matching Symptoms to Faults

1. **Monitor Operating Parameters** – Track pressure, temperature, and flow rates regularly to identify deviations.
2. **Inspect Visually** – Regularly inspect the system for visible leaks, corrosion, or other signs of wear.
3. **Use Diagnostic Tools** – Employ tools like ultrasonic detectors, thermal cameras, and pressure gauges to detect hidden issues.
4. **Check Maintenance Logs** – Reviewing maintenance history can help identify recurring issues related to certain faults.
5. **Consult Manufacturers' Guidelines** – Refer to manuals and specifications to ensure components are functioning within their recommended parameters.

16.6.4 Summary

In summary, understanding the connection between symptoms and faults is essential for effective troubleshooting and repair in piping systems, preventing costly downtime, and ensuring smooth operations.

CHAPTER SEVENTEEN

17.0 MAINTENANCE PROCEDURES FOR CURRENT ENGINE DESIGNS

Maintenance procedures for current engine designs, whether diesel or gasoline, are crucial to ensure long-lasting performance, fuel efficiency, and safety. These procedures involve regular checks, servicing, and component replacements.

17.1 KEY MAINTENANCE TASKS – ROUTINE MAINTENANCE AND SCHEDULED SERVICING FOR BOTH DIESEL AND GASOLINE ENGINES

17.1.1 Routine Maintenance

Routine maintenance refers to simple tasks that can be performed frequently to keep the engine in good condition.

a) Checking and Changing Engine Oil

- **Frequency** – Every 5,000 to 10,000 miles (8,000 to 16,000 km) or as recommended by the manufacturer.
- **Procedure**

1 Warm the engine for a few minutes to circulate the oil.

2 Turn off the engine and locate the oil drain plug underneath the vehicle.

3 Place an oil catch pan beneath the plug and remove it to drain old oil.

4 Remove the old oil filter and replace it with a new one (apply a small amount of new oil to the new filter's seal before installation).

5 Reinstall the oil drain plug and refill the engine with the manufacturer-recommended type and quantity of oil.

6 Check the oil level with the dipstick to ensure it is correct.

- **Tips**
  - Always use the correct grade and type of oil recommended for the engine.
  - Dispose of used oil and filters at a recycling facility.

b) Inspecting and Replacing the Air Filter

- **Frequency** – Every 12,000 to 15,000 miles (19,000 to 24,000 km) or annually.
- **Procedure**

1 Locate the air filter housing, usually in the engine bay.

2 Remove the housing cover to access the air filter.

3 Take out the old filter and check for dirt or debris.

4 Replace the filter if it is dirty or damaged.

5 Reassemble the housing.

- **Tips**
  - A clean air filter improves fuel efficiency and engine performance.
  - Check the air intake for any blockages.

c) Checking and Topping Up Fluids

- **Frequency** – Monthly or every 1,000 miles (1,600 km).
- **Procedure**

1 **Coolant** – Check the coolant reservoir and ensure the level is between the minimum and maximum marks. Add a 50/50 mix of coolant and water if necessary.

2 **Brake Fluid** – Inspect the brake fluid reservoir and top up with the manufacturer-recommended brake fluid if it's low.

3 **Power Steering Fluid** – Check the power steering reservoir and add fluid if needed.

4 **Transmission Fluid** – For automatic transmissions, check the dipstick (if applicable) for correct fluid levels. Top up with the recommended type if necessary.

- **Tips**
  - Use only the recommended fluids for your specific engine type.
  - Never open the radiator cap when the engine is hot—use the coolant reservoir instead.

d) Inspecting Belts and Hoses

- **Frequency** – Every 15,000 miles (24,000 km) or as needed.
- **Procedure**

1 Check the serpentine belt for cracks, fraying, or glazing.

2 Inspect hoses for leaks, cracks, or hardening.

3 Replace any worn or damaged belts or hoses.

- **Tips**
  - A squealing noise during engine operation can indicate a loose or worn belt.
  - Hoses should feel firm, not brittle or soft.

17.1.2 Scheduled Servicing

Scheduled servicing involves more thorough maintenance tasks that should be performed at specific intervals, as indicated by the vehicle manufacturer.

a) Replacing the Fuel Filter

- **Frequency** – Every 20,000 to 40,000 miles (32,000 to 64,000 km) for gasoline engines and 10,000 to 15,000 miles (16,000 to 24,000 km) for diesel engines.
- **Procedure**

1 Relieve the fuel pressure by removing the fuel pump fuse or relay and cranking the engine.

2 Locate the fuel filter (often along the fuel line or near the fuel tank).

3 Remove the old filter, paying attention to fuel spillage.

4 Install the new filter, ensuring proper flow direction.

5 Reassemble and test the fuel system.

- **Tips**
  - A clean fuel filter prevents clogging and ensures proper fuel delivery.
  - Wear protective gloves and eyewear when handling fuel.

b) Inspecting and Replacing Spark Plugs (Gasoline Engines)

- **Frequency** – Every 30,000 to 60,000 miles (48,000 to 96,000 km), depending on the plug type.
- **Procedure**

1 Disconnect the battery's negative terminal.

2 Remove the ignition coil or spark plug wires to access the spark plugs.

3 Use a spark plug socket to remove the old spark plugs.

4 Check the gap on the new plugs and adjust if necessary.

5 Install the new spark plugs and torque to the manufacturer's specifications.

6 Reconnect the ignition components and battery.

- **Tips**
  - Always use the correct type of spark plug recommended for your engine.
  - A misfire or rough idle could indicate worn spark plugs.

c) Replacing the Timing Belt (If Applicable)

- **Frequency** – Every 60,000 to 100,000 miles (96,000 to 160,000 km) for both diesel and gasoline engines.
- **Procedure**

1 Remove components obstructing access to the timing belt, such as the serpentine belt, covers, or accessories.

2 Align the camshaft and crankshaft to the correct timing marks.

3 Loosen the tensioner and remove the old timing belt.

4 Install the new timing belt, maintaining the correct alignment of timing marks.

5 Re-tension the belt and reassemble the engine components.

6 Test the engine for proper operation.

- **Tips**
  - Replace the water pump and tensioner when doing a timing belt change.
  - A broken timing belt can cause severe engine damage, especially in interference engines.

d) Inspecting and Replacing the Battery

- **Frequency** – Every 3 to 5 years or as needed.
- **Procedure**

1 Inspect the battery terminals for corrosion. Clean if necessary.

2 Check the battery voltage using a multimeter; it should read around 12.6 volts when the engine is off.

3 If the battery is weak, disconnect the negative terminal first, followed by the positive.

4 Replace the old battery with a new one and reconnect the terminals, starting with the positive terminal.

- **Tips**
  - Test the alternator to ensure it's charging the battery properly.
  - Dispose of the old battery at a recycling facility.

e) Inspecting the Exhaust System

- **Frequency** – Annually or every 20,000 to 30,000 miles (32,000 to 48,000 km).
- **Procedure**

1 Check the exhaust pipes, muffler, and catalytic converter for leaks, rust, or damage.

2 Listen for unusual noises like hissing or rattling.

3 Replace any damaged or corroded parts.

- **Tips**
  - A properly functioning exhaust system helps reduce emissions and maintain engine performance.
  - Use exhaust sealant for minor leaks if replacement isn't necessary.

f) Cleaning and Inspecting the EGR Valve (Diesel Engines)

- **Frequency** – Every 30,000 miles (48,000 km) or as recommended.
- **Procedure**

1 Disconnect the battery.

2 Remove the EGR valve from the engine.

3 Use an EGR valve cleaner to remove carbon buildup.

4 Reinstall the cleaned valve and reconnect the battery.

- **Tips**
  - A clogged EGR valve can cause poor fuel economy and increased emissions.
  - Use only recommended cleaners for the EGR system.

17.1.3 General Maintenance Tips

- **Keep an Eye on Warning Lights** – Always address any illuminated dashboard warning lights promptly to avoid potential damage.
- **Check Tire Pressure Regularly** – Proper tire inflation improves fuel economy and handling.

- **Test for Diagnostic Codes** – Use an OBD-II scanner to read diagnostic trouble codes (DTCs) if the Check Engine Light is on.
- **Schedule Professional Inspections** – Even if you perform regular maintenance, a professional mechanic should inspect the vehicle at least once a year.

17.1.4 Summary

Proper maintenance of diesel and gasoline engines involves a combination of routine checks and scheduled servicing. Adhering to the manufacturer's maintenance schedule ensures optimal engine performance, fuel efficiency, and longevity. Always refer to the vehicle's manual for specific maintenance intervals and recommended products for your engine type.

17.2 MATCHING SYMPTOMS AND FAULTS IN MODERN ENGINES, BOTH DIESEL AND GASOLINE

Matching symptoms and faults in modern engines, both diesel and gasoline, is crucial for effective diagnostics and repair.

17.2.1 Common Symptoms and the Corresponding Faults for Both Diesel and Gasoline Engines

1 Hard Starting or No Start

- **Symptoms**
  - Engine struggles to start.
  - Cranking takes longer than usual.
  - No response when turning the key.
- **Potential Faults**
  - **Battery Issues** – Weak or dead battery, loose battery connections.
  - **Fuel System Problems** – Clogged fuel filter, fuel pump failure, or lack of fuel pressure.
  - **Ignition System Failure (Gasoline)** – Bad spark plugs, ignition coil failure, or faulty crankshaft position sensor.
  - **Compression Issues (Diesel)** – Worn piston rings, cylinder wear, or valve timing problems.

2 **Engine Misfiring**

- **Symptoms**
  - Rough idling.
  - Lack of power or hesitation during acceleration.
  - Engine shaking or vibrations.
  - Illuminated Check Engine Light (CEL).
- **Potential Faults**
  - **Ignition Problems (Gasoline)** – Faulty spark plugs, bad ignition coils, or damaged spark plug wires.
  - **Fuel Injector Issues** – Clogged or leaking fuel injectors in both diesel and gasoline engines.
  - **Vacuum Leak** – Broken or loose vacuum hoses, intake manifold gasket leaks.
  - **Air-Fuel Mixture Imbalance** – Faulty oxygen sensors, mass air flow sensor issues.

3 **Excessive Exhaust Smoke**

- **Symptoms**
  - Black, white, or blue smoke from the exhaust.
  - Strong fuel odor.
- **Potential Faults**
  - **Black Smoke (Diesel & Gasoline)** – Overfueling due to a clogged air filter, bad fuel injectors, faulty turbocharger, or EGR (Exhaust Gas Recirculation) valve issues.
  - **White Smoke** – Coolant leakage into the combustion chamber, usually from a blown head gasket, cracked cylinder head, or damaged engine block.
  - **Blue Smoke** – Burning oil, likely due to worn piston rings, valve seals, or turbocharger issues.

4 **Poor Fuel Economy**

- **Symptoms**
  - Increased fuel consumption.
  - Frequent refuelling.
  - Reduced mileage.

- **Potential Faults**
  - **Dirty Air Filter** – Restricts air intake, causing the engine to burn more fuel.
  - **Faulty Oxygen Sensors (Gasoline)** – Incorrect air-fuel mixture readings.
  - **Leaking Fuel Injectors** – Fuel leaking into the combustion chamber.
  - **Low Tire Pressure or Wheel Alignment Issues** – Increased rolling resistance impacting fuel efficiency.

5 **Engine Overheating**

- **Symptoms**
  - High temperature gauge reading.
  - Steam from the engine bay.
  - Warning light for overheating.
- **Potential Faults**
  - **Low Coolant Levels** – Leaks in the cooling system or insufficient coolant.
  - **Faulty Thermostat** – Stuck closed, preventing coolant circulation.
  - **Water Pump Failure** – Unable to circulate coolant effectively.
  - **Radiator Issues** – Clogged radiator or broken cooling fan.

6 **Loss of Power or Poor Acceleration**

- **Symptoms**
  - Lack of engine responsiveness.
  - Slow acceleration or loss of power under load.
  - Trouble reaching high speeds.
- **Potential Faults**
  - **Clogged Fuel Filter** – Restricts fuel flow to the engine.
  - **Throttle Body Problems (Gasoline)** – Dirty or malfunctioning throttle body affecting air intake.
  - **Turbocharger Failure (Diesel)** – Reduced boost pressure due to damaged or leaking turbocharger.
  - **Transmission Issues** – Slipping transmission or issues with the clutch (manual transmission).

7 **Unusual Engine Noises**

- **Symptoms**
  - Knocking, pinging, or rattling sounds.
  - Hissing or whining noises.
  - Grinding or squealing sounds.
- **Potential Faults**
  - **Engine Knock** – Low-quality fuel causing pre-detonation, or carbon buildup on pistons (common in gasoline engines).
  - **Loose Timing Chain/Belt** – Causing rattling or ticking noises.
  - **Accessory Belt Issues** – Squealing from worn or loose serpentine belt.
  - **Exhaust Leak** – Hissing noise from a crack or hole in the exhaust system.
  - **Bad Bearings** – Worn bearings in components like the alternator, water pump, or tensioners.

8 **Rough Idling**

- **Symptoms**
  - Shaky or rough engine when idling.
  - Fluctuating RPMs when the vehicle is stationary.
- **Potential Faults**
  - **Dirty Throttle Body or Idle Air Control Valve (Gasoline)**: Inconsistent air flow affecting idle.
  - **Vacuum Leaks** – Unmetered air entering the engine.
  - **Fuel Injector Problems** – Inconsistent fuel delivery.
  - **Spark Plug or Ignition Coil Issues (Gasoline)** – Misfiring during idle.

9 **Transmission Shifting Issues**

- **Symptoms**
  - Delayed or hard shifts.
  - Slipping gears.
  - Noisy shifts.

- **Potential Faults**
  - **Low Transmission Fluid** – Causing poor lubrication and shifting difficulties.
  - **Worn Clutch (Manual Transmission)** – Slipping and poor power transfer.
  - **Transmission Control Module Faults** – Erratic or incorrect shifting behavior.
  - **Torque Converter Issues (Automatic)** – Causing slippage or shudder.

10 Illuminated Check Engine Light (CEL)

- **Symptoms**
  - Check Engine Light is illuminated or flashing.
  - Other warning lights may also be on (e.g., ABS, Traction Control).
- **Potential Faults**
  - **Faulty Sensors** – Malfunctioning oxygen sensor, mass air flow sensor, or camshaft/crankshaft sensors.
  - **Catalytic Converter Issues** – Inefficient catalytic converter performance.
  - **EVAP System Leak** – Leaks in the Evaporative Emission Control System.
  - **General Engine Issues** – Misfires, exhaust leaks, or other minor engine faults.

11 Unresponsive Throttle or Stalling

- **Symptoms**
  - Throttle delay or unresponsive acceleration.
  - Stalling at idle or during driving.
- **Potential Faults**
  - **Faulty Throttle Position Sensor (Gasoline)** – Incorrect throttle readings.
  - **Dirty Mass Air Flow Sensor** – Affects air-fuel mixture.
  - **EGR Valve Problems** – Sticking or blocked EGR valve leading to stalling.
  - **Fuel Pump Failure** – Intermittent fuel delivery causing stalling.

12 Oil Leaks

- **Symptoms**
  - Oil spots under the car.
  - Burning oil smell or smoke.

- Low oil level warning.

- **Potential Faults**
  - **Worn Gaskets or Seals** – Oil pan gasket, valve cover gasket, or front/rear main seal.
  - **Oil Filter Issues** – Improperly installed or damaged oil filter.
  - **Loose Drain Plug** – Causing oil to leak from the oil pan.

17.2.2 Summary

Diagnosing faults in modern diesel and gasoline engines requires understanding both visual and auditory symptoms and using diagnostic tools like an OBD-II scanner for precise fault codes. Regular maintenance and keeping components clean can prevent many of these common issues, leading to a longer engine lifespan and fewer unexpected breakdowns.

17.3 MATCHING SYMPTOMS TO FAULTS IN MODERN ENGINE DESIGNS

Matching symptoms to faults in modern engine designs is essential for diagnosing issues and preventing further damage.

17.3.1 Common Symptoms in Modern Gasoline and Diesel Engines and Their Likely Causes

1 **Hard Starting or No Start**

- **Symptoms**
  - Engine struggles to start.
  - Cranking takes longer than usual.
  - No response when turning the key.
- **Potential Faults**
  - **Battery Issues** – Weak or dead battery, loose battery connections.
  - **Fuel System Problems** – Clogged fuel filter, fuel pump failure, or lack of fuel pressure.
  - **Ignition System Failure (Gasoline)** – Bad spark plugs, ignition coil failure, or faulty crankshaft position sensor.
  - **Compression Issues (Diesel)** – Worn piston rings, cylinder wear, or valve timing problems.

2 **Engine Misfiring**

- **Symptoms**
  - Rough idling.
  - Lack of power or hesitation during acceleration.
  - Engine shaking or vibrations.
  - Illuminated Check Engine Light (CEL).
- **Potential Faults**
  - **Ignition Problems (Gasoline)** – Faulty spark plugs, bad ignition coils, or damaged spark plug wires.
  - **Fuel Injector Issues** – Clogged or leaking fuel injectors in both diesel and gasoline engines.
  - **Vacuum Leak** – Broken or loose vacuum hoses, intake manifold gasket leaks.
  - **Air-Fuel Mixture Imbalance** – Faulty oxygen sensors, mass air flow sensor issues.

3 **Excessive Exhaust Smoke**

- **Symptoms**
  - Black, white, or blue smoke from the exhaust.
  - Strong fuel odor.
- **Potential Faults**
  - **Black Smoke (Diesel & Gasoline)** – Overfueling due to a clogged air filter, bad fuel injectors, faulty turbocharger, or EGR (Exhaust Gas Recirculation) valve issues.
  - **White Smoke** – Coolant leakage into the combustion chamber, usually from a blown head gasket, cracked cylinder head, or damaged engine block.
  - **Blue Smoke** – Burning oil, likely due to worn piston rings, valve seals, or turbocharger issues.

4 **Poor Fuel Economy**

- **Symptoms**
  - Increased fuel consumption.
  - Frequent refueling.
  - Reduced mileage.

- **Potential Faults**
  - **Dirty Air Filter** – Restricts air intake, causing the engine to burn more fuel.
  - **Faulty Oxygen Sensors (Gasoline)** – Incorrect air-fuel mixture readings.
  - **Leaking Fuel Injectors** – Fuel leaking into the combustion chamber.
  - **Low Tire Pressure or Wheel Alignment Issues** – Increased rolling resistance influencing fuel efficiency.

5 **Engine Overheating**

- **Symptoms**
  - High temperature gauge reading.
  - Steam from the engine bay.
  - Warning light for overheating.
- **Potential Faults**
  - **Low Coolant Levels** – Leaks in the cooling system or insufficient coolant.
  - **Faulty Thermostat** – Stuck closed, preventing coolant circulation.
  - **Water Pump Failure** – Unable to circulate coolant effectively.
  - **Radiator Issues** – Clogged radiator or broken cooling fan.

6 **Loss of Power or Poor Acceleration**

- **Symptoms**
  - Lack of engine responsiveness.
  - Slow acceleration or loss of power under load.
  - Trouble reaching high speeds.
- **Potential Faults**
  - **Clogged Fuel Filter** – Restricts fuel flow to the engine.
  - **Throttle Body Problems (Gasoline)** – Dirty or malfunctioning throttle body affecting air intake.
  - **Turbocharger Failure (Diesel)** – Reduced boost pressure due to damaged or leaking turbocharger.
  - **Transmission Issues** – Slipping transmission or issues with the clutch (manual transmission).

7 **Unusual Engine Noises**

- **Symptoms**
    - Knocking, pinging, or rattling sounds.
    - Hissing or whining noises.
    - Grinding or squealing sounds.
- **Potential Faults**
    - **Engine Knock** – Low-quality fuel causing pre-detonation, or carbon buildup on pistons (common in gasoline engines).
    - **Loose Timing Chain/Belt** – Causing rattling or ticking noises.
    - **Accessory Belt Issues** – Squealing from worn or loose serpentine belt.
    - **Exhaust Leak** – Hissing noise from a crack or hole in the exhaust system.
    - **Bad Bearings** – Worn bearings in components like the alternator, water pump, or tensioners.

8 **Rough Idling**

- **Symptoms**
    - Shaky or rough engine when idling.
    - Fluctuating RPMs when the vehicle is stationary.
- **Potential Faults**
    - **Dirty Throttle Body or Idle Air Control Valve (Gasoline)** – Inconsistent air flow affecting idle.
    - **Vacuum Leaks** – Unmetered air entering the engine.
    - **Fuel Injector Problems** – Inconsistent fuel delivery.
    - **Spark Plug or Ignition Coil Issues (Gasoline)**: Misfiring during idle.

9 **Transmission Shifting Issues**

- **Symptoms**
    - Delayed or hard shifts.
    - Slipping gears.
    - Noisy shifts.

- **Potential Faults**
  - **Low Transmission Fluid** – Causing poor lubrication and shifting difficulties.
  - **Worn Clutch (Manual Transmission)** – Slipping and poor power transfer.
  - **Transmission Control Module Faults** – Erratic or incorrect shifting behavior.
  - **Torque Converter Issues (Automatic)** – Causing slippage or shudder.

10 Illuminated Check Engine Light (CEL)

- **Symptoms**
  - Check Engine Light is illuminated or flashing.
  - Other warning lights may also be on (e.g., ABS, Traction Control).
- **Potential Faults**
  - **Faulty Sensors** – Malfunctioning oxygen sensor, mass air flow sensor, or camshaft/crankshaft sensors.
  - **Catalytic Converter Issues** – Inefficient catalytic converter performance.
  - **EVAP System Leak** – Leaks in the Evaporative Emission Control System.
  - **General Engine Issues** – Misfires, exhaust leaks, or other minor engine faults.

11 Unresponsive Throttle or Stalling

- **Symptoms**
  - Throttle delay or unresponsive acceleration.
  - Stalling at idle or during driving.
- **Potential Faults**
  - **Faulty Throttle Position Sensor (Gasoline)** – Incorrect throttle readings.
  - **Dirty Mass Air Flow Sensor** – Affects air-fuel mixture.
  - **EGR Valve Problems** – Sticking or blocked EGR valve leading to stalling.
  - **Fuel Pump Failure** – Intermittent fuel delivery causing stalling.

12 Oil Leaks

- **Symptoms**
  - Oil spots under the car.
  - Burning oil smell or smoke.

- Low oil level warning.

- **Potential Faults**
  - **Worn Gaskets or Seals** – Oil pan gasket, valve cover gasket, or front/rear main seal.
  - **Oil Filter Issues** – Improperly installed or damaged oil filter.
  - **Loose Drain Plug** – Causing oil to leak from the oil pan.

17.3.2 Summary

Diagnosing faults in modern diesel and gasoline engines requires understanding both **visual and auditory symptoms** and using diagnostic tools like an OBD-II scanner for precise fault codes. Regular maintenance and keeping components clean can prevent many of these common issues, leading to a longer engine lifespan and fewer unexpected breakdowns.

17.4 DIAGNOSING FAULTS IN CURRENT DIESEL ENGINES

When diagnosing faults in current diesel engines, recognizing the symptoms is crucial for accurate troubleshooting. This section considers a detailed guide on common symptoms in modern diesel engines and the likely faults that could be causing them.

17.4.1 Detailed Guide on Common Symptoms in Modern Diesel Engines

1 Symptom: Engine Misfire or Rough Idle

- **Possible Faults**
  - **Ignition System Fault** – Worn-out spark plugs, faulty ignition coils, or malfunctioning spark plug wires can lead to misfiring in gasoline engines.
  - **Fuel Injector Issues** – Clogged or malfunctioning fuel injectors (in both gasoline and diesel engines) may cause improper fuel delivery, leading to misfires.
  - **Vacuum Leaks** – Air leaks in the intake manifold or vacuum hoses can disrupt the air-fuel mixture, causing misfires.
  - **Faulty Engine Control Unit (ECU)** – A malfunctioning ECU may improperly manage fuel and spark timing, leading to engine misfires.
  - **Low Compression** – Worn-out piston rings, valves, or head gaskets can lead to low compression, causing misfires in diesel engines.

2 **Symptom: Excessive Smoke from Exhaust**

- **Possible Faults**
  - **Black Smoke** (Excess Fuel)
    - **Fuel System Fault** – In diesel engines, a clogged air filter, malfunctioning fuel injectors, or faulty fuel pressure regulator can lead to an overly rich air-fuel mixture, causing black smoke.
    - **Dirty Air Filter** – Restricted airflow due to a clogged air filter can result in incomplete combustion, leading to black smoke.
  - **Blue Smoke** (Oil Consumption)
    - **Worn Valve Seals or Piston Rings** – If the valve seals or piston rings are worn, oil can enter the combustion chamber, burning and causing blue smoke, common in both diesel and gasoline engines.
  - **White Smoke** (Coolant in Combustion Chamber)
    - **Blown Head Gasket** – A blown head gasket can cause coolant to enter the combustion chamber, leading to white smoke in both diesel and gasoline engines.
    - **Faulty EGR Valve** – In diesel engines, a malfunctioning Exhaust Gas Recirculation (EGR) valve can cause incomplete combustion, potentially leading to white smoke.

3 **Symptom: Engine Overheating**

- **Possible Faults**
  - **Low Coolant Level** – A common cause of engine overheating is simply low coolant levels, due to leaks or evaporation.
  - **Thermostat Malfunction** – If the thermostat is stuck closed, it prevents coolant from circulating properly, leading to overheating.
  - **Faulty Radiator** – Blocked or leaking radiators can hinder heat dissipation, causing the engine to overheat.
  - **Water Pump Failure** – If the water pump fails or its belt breaks, coolant circulation stops, leading to engine overheating.
  - **Cooling Fan Failure** – A malfunctioning cooling fan can prevent adequate air circulation across the radiator, causing overheating, especially in heavy traffic conditions.

4 **Symptom: Low Engine Power or Poor Acceleration**

- **Possible Faults**
  - **Fuel System Issues** – Clogged fuel injectors or a faulty fuel pump can prevent the engine from receiving adequate fuel, leading to poor acceleration or a loss of power.
  - **Air Filter Clogging** – A clogged air filter restricts airflow to the engine, leading to reduced power and poor performance.
  - **Turbocharger Issues (Diesel)** – If the turbocharger or its components (e.g., intercooler, wastegate) malfunction, it can reduce engine power significantly, especially in diesel engines.
  - **Exhaust Restrictions** – A clogged Diesel Particulate Filter (DPF) or exhaust system restrictions can reduce engine power in diesel engines, leading to a noticeable loss of acceleration.
  - **Timing Problems** – Incorrect ignition timing or a faulty timing belt/chain can lead to poor engine performance and hesitation during acceleration.

5 **Symptom: Rough Idle**

- **Possible Faults**
  - **Vacuum Leaks** – Leaks in the vacuum system can cause an uneven idle by disrupting the air-fuel mixture.
  - **Dirty or Faulty Fuel Injectors** – Malfunctioning or dirty fuel injectors can cause an uneven fuel delivery, resulting in rough idling.
  - **Faulty Idle Air Control Valve (IAC)** – The IAC valve, which controls idle speed, can malfunction and cause the engine to idle roughly.
  - **Mass Air Flow Sensor (MAF) Issues** – A malfunctioning MAF sensor can send incorrect airflow data to the ECU, causing improper fuel mixture and rough idle.

6 **Symptom: Increased Fuel Consumption**

- **Possible Faults**
  - **Oxygen Sensor Malfunction** – A faulty oxygen sensor can cause the engine to run rich (too much fuel), leading to increased fuel consumption.
  - **Dirty or Clogged Air Filter** – A clogged air filter reduces engine efficiency by restricting airflow, which forces the engine to use more fuel.

- **Faulty Fuel Injectors** – Leaking or clogged fuel injectors can result in an incorrect fuel mixture, causing higher fuel consumption.
- **Faulty Thermostat** – A stuck-open thermostat can prevent the engine from reaching the correct operating temperature, causing it to run inefficiently and consume more fuel.
- **Low Tire Pressure** – Underinflated tires can cause increased rolling resistance, which results in more fuel being used to maintain speed.

7 **Symptom: Engine Knocking or Pinging**

- **Possible Faults**
  - **Incorrect Fuel Octane (Gasoline)** – Using fuel with an octane rating lower than the engine's specifications can cause knocking or pinging, especially under load.
  - **Carbon Deposits** – Excessive carbon buildup in the combustion chamber can increase compression, leading to knocking, particularly in diesel engines.
  - **Faulty Knock Sensor** – A malfunctioning knock sensor may fail to detect abnormal combustion, causing the engine to knock.
  - **Advanced Ignition Timing** – If the ignition timing is set too advanced, it can cause the air-fuel mixture to ignite prematurely, resulting in knocking.

8 **Symptom: Check Engine Light (CEL) On**

- **Possible Faults**
  - **Faulty Oxygen Sensor** – A faulty oxygen sensor is a common trigger for the check engine light to turn on, as it affects the engine's air-fuel mixture.
  - **Mass Air Flow (MAF) Sensor Issue** – A malfunctioning MAF sensor can also trigger the check engine light, as it misleads the engine control unit about the air entering the engine.
  - **Exhaust Gas Recirculation (EGR) System Fault** – Issues with the EGR valve, such as clogging or malfunctioning, can also cause the check engine light to illuminate.
  - **Fuel System Faults** – Clogged fuel injectors or a failing fuel pump can lead to the CEL being triggered due to poor fuel delivery.

9 **Symptom: Hard Starting or No Start**

- **Possible Faults**
  - **Weak or Dead Battery** – If the battery is weak or dead, it will not provide enough power to start the engine.

- **Starter Motor Failure** – A faulty starter motor may prevent the engine from cranking.
- **Fuel Delivery Issues** – Problems with the fuel pump, fuel filter, or fuel injectors can cause inadequate fuel delivery, preventing the engine from starting.
- **Glow Plug Problems (Diesel)** – In diesel engines, faulty glow plugs can prevent the engine from starting, particularly in cold weather.

10 Symptom: Excessive Vibrations

- **Possible Faults**
  - **Engine Mount Problems** – Worn or damaged engine mounts can cause excessive engine movement, resulting in vibrations felt throughout the vehicle.
  - **Unbalanced Components** – Issues with the crankshaft, flywheel, or harmonic balancer can cause vibrations.
  - **Injector Issues (Diesel)** – Faulty or unbalanced fuel injectors in diesel engines can cause vibrations due to irregular fuel delivery.

17.4.2 Summary

Matching symptoms with faults requires careful observation of the engine's behavior and, in many cases, using diagnostic tools such as an OBD-II scanner or a multimeter. Routine maintenance, such as changing filters, checking fluid levels, and ensuring proper timing, can prevent many of these issues. Regular inspections and early detection are key to preventing engine damage and ensuring the vehicle operates at peak performance. Always refer to the manufacturer's manual for specific maintenance intervals and fault diagnostics for the particular engine model in question.

17.5 REPAIRING SIMPLE FAULTS AND REPLACING COMPONENTS IN MODERN ENGINES

Repairing simple faults and replacing components in modern engines (both diesel and gasoline) requires careful attention to detail, proper tools, and knowledge of engine systems.

17.5.1 Step-By-Step Guide for Repairing Common Faults and Replacing Parts in Current Engine Designs

1 Fault: Clogged or Faulty Air Filter

Symptoms – Reduced engine power, poor acceleration, rough idling, black smoke.

Repair Steps

a) **Locate the Air Filter** – The air filter is usually located in the air intake system, secured in a housing.

b) **Remove the Housing** – Unscrew or unclip the air filter housing to access the filter.

c) **Inspect the Air Filter** – Check for dirt, debris, or clogging. If the filter is dirty or damaged, it needs to be replaced.

d) **Replace the Air Filter** – Remove the old filter and replace it with a new one. Ensure it is seated properly in the housing.

e) **Reassemble the Housing** – Close the housing and tighten any screws or clips.

Tips

- Always use a manufacturer-recommended air filter.
- Inspect the air intake for any damage or blockages.

2 Fault: Faulty Fuel Injectors

Symptoms – Misfiring, rough idle, loss of power, excessive smoke from exhaust, poor fuel efficiency.

Repair Steps

a) **Identify the Faulty Injector** – Use a diagnostic tool (OBD scanner) to identify the faulty injector or conduct a fuel injector flow test.

b) **Disconnect the Battery** – To avoid electrical shocks, disconnect the negative terminal of the battery.

c) **Access the Injectors** – Remove any components obstructing access to the injectors, such as the intake manifold or engine cover.

d) **Remove the Faulty Injector** – Unscrew or unclamp the injector from its holder. Be careful with the fuel lines connected to the injector.

e) **Install New Injector** – Insert the new fuel injector, ensuring it is seated properly. Tighten any retaining bolts or clips.

f) **Reconnect Fuel Lines and Electrical Connectors** – Attach the fuel line and electrical connectors back to the new injector.

g) **Reassemble the Engine** – Put back any parts that were removed, such as the intake manifold or engine cover.

h) **Test the Engine** – Reconnect the battery and start the engine to check if the misfire or rough idle has been resolved.

Tips

- Always replace injectors in pairs or all cylinders if one fails.
- Clean the fuel system if necessary to prevent clogging.

3 Fault: Dead or Weak Battery

Symptoms – Difficulty starting, dim headlights, electrical accessories not functioning properly.

Repair Steps

a) **Check the Battery Voltage** – Use a multimeter to check the battery's voltage. A healthy battery should read around 12.6V when the engine is off.

b) **Inspect Battery Terminals** – Check for corrosion or loose connections at the battery terminals. Clean and tighten the terminals if necessary.

c) **Test the Alternator** – If the battery is charged but still weak, the alternator may not be charging it properly. Test the alternator's output using a multimeter (it should read around 14V with the engine running).

d) **Replace the Battery**

- Disconnect the negative terminal first, followed by the positive terminal.
- Remove the battery hold-down bracket and lift the battery out of the tray.
- Place the new battery in the tray and secure it with the bracket.
- Reconnect the positive terminal first, then the negative terminal.

e) **Test the System** – Start the engine and check for proper operation. The vehicle should start easily, and electrical systems should function normally.

Tips

- Clean corrosion off terminals with a wire brush or a mixture of baking soda and water.
- Dispose of old batteries properly at a recycling facility.

4 Fault: Clogged Fuel Filter

Symptoms – Engine stalling, rough idle, poor fuel efficiency, hard starting, loss of power.

Repair Steps

a) **Locate the Fuel Filter** – The fuel filter is typically located along the fuel line between the fuel tank and the engine. It may be in the engine bay or underneath the vehicle.

b) **Relieve Fuel Pressure** – Before removing the fuel filter, relieve fuel pressure by disconnecting the fuel pump relay or fuse and cranking the engine to release pressure.

c) **Disconnect the Fuel Lines** – Using the appropriate tools, disconnect the fuel lines from the filter. Be prepared for small amounts of fuel to leak out.

d) **Remove the Old Fuel Filter** – Unscrew or unclip the filter from its mount. Take note of the filter's orientation for proper installation.

e) **Install the New Fuel Filter** – Position the new filter in the same orientation as the old one. Secure it in place and reconnect the fuel lines.

f) **Test the Fuel System** – Reconnect the fuel pump relay or fuse, then start the engine to check for leaks and proper fuel flow.

Tips

- Replace the fuel filter at manufacturer-recommended intervals to maintain engine performance.
- Be cautious when working with fuel lines to avoid fuel spillage and fire hazards.

5 Fault: Worn or Broken Serpentine Belt

Symptoms – Squealing noise from the engine, failure of accessories such as alternator, power steering, or air conditioning.

Repair Steps

a) **Locate the Serpentine Belt** – The serpentine belt is located on the front of the engine and drives various accessories.

b) **Inspect the Belt** – Check for signs of wear, such as cracks, fraying, or glazing. If the belt is damaged or worn, it needs to be replaced.

c) **Release Tension** – Use a serpentine belt tool or a wrench to relieve tension from the belt by rotating the tensioner pulley.

d) **Remove the Old Belt** – Once tension is released, slide the old belt off the pulleys.

e) **Install the New Belt** – Compare the new belt with the old one to ensure proper fit. Route the new belt over the pulleys as per the manufacturer's diagram (usually located on the engine).

f) **Reapply Tension** – Reapply tension to the belt by releasing the tensioner pulley.

g) **Test the Engine** – Start the engine and listen for any unusual noises or vibrations, ensuring that all accessories are functioning properly.

Tips

- Ensure the new belt is properly aligned on all pulleys.
- Inspect pulleys for wear or damage when replacing the belt.

6 Fault: Worn or Broken Timing Belt (for engines that use a timing belt)

Symptoms – Engine misfires, difficulty starting, or a "slapping" sound from the timing belt area.

Repair Steps

a) **Disconnect the Battery** – Disconnect the negative terminal to avoid electrical issues.

b) **Remove the Timing Belt Covers** – You may need to remove components like the alternator or radiator to access the timing belt.

c) **Align the Timing Marks** – Before removing the old timing belt, rotate the engine to align the timing marks on the camshaft and crankshaft pulleys.

d) **Remove the Old Timing Belt** – Loosen the tensioner and remove the belt from the pulleys.

e) **Install the New Timing Belt** – Place the new belt on the pulleys, ensuring the timing marks align. Tension the belt as per the manufacturer's specifications.

f) **Reassemble the Engine** – Reattach the timing belt covers and any other components you removed.

g) **Test the Engine** – Start the engine and check for smooth operation.

Tips

- Replace the timing belt at recommended intervals (typically every 60,000 to 100,000 miles).
- Check the tensioner, pulleys, and water pump for wear while replacing the timing belt.

17.5.1 Summary

Repairing simple faults and replacing components in current engine designs requires basic mechanical knowledge, proper tools, and safety precautions. It is important to always follow the manufacturer's service manual for specific procedures, torque specifications, and parts recommendations. If you're unfamiliar with a repair, it's always safer to consult a professional mechanic. Regular maintenance can prevent many of these faults from occurring in the first place, ensuring the engine runs smoothly and efficiently.

CHAPTER EIGHTEEN

18.0 MAINTENANCE, TESTING, AND REPAIR OF ELECTRICAL INSTALLATIONS, INSTRUMENTS, AND EQUIPMENT

18.1 ELECTRICAL INSTALLATIONS

Electrical installations refer to the complete system of electrical wiring, equipment, and components that provide power and control for a building or facility. This includes the installation of electrical circuits, outlets, switches, distribution boards, lighting systems, and other devices necessary for the safe and efficient operation of electrical systems. Electrical installations must adhere to local and national safety standards, ensuring proper grounding, insulation, and protection against overloads and faults. These installations are designed to meet the specific power requirements of the building or space, while prioritizing safety, reliability, and energy efficiency. Proper installation and maintenance are essential to prevent hazards like electrical fires, shocks, or system failures.

18.1.1 General Description of Space Structure in Electrical Installation

The space structure in electrical installation refers to the arrangement and organization of electrical components and systems within a given physical space. It encompasses the layout, distribution, and management of wiring, equipment, and devices such as circuit breakers, switches, outlets, and control panels. The space structure is crucial for ensuring safety, efficiency, and functionality of the electrical system. Proper planning and design of the space structure allow for easy maintenance, future upgrades, and compliance with electrical codes and standards, while also optimizing the use of available space within residential, commercial, or industrial settings.

a) Workbenches and Stations

- Central or wall-mounted workbenches equipped with tools for assembly, repair, and testing.
- Specialized stations for tasks like soldering, wiring, and testing electrical circuits.
- Each workstation usually includes accessible storage for frequently used tools, such as pliers, screwdrivers, wire strippers, soldering irons, and multimeters.

b) Tool and Component Storage

- Cabinets or drawers for organizing hand tools, small parts, and specialty tools.
- Shelving units for larger tools like drills, power supplies, and other equipment.
- Transparent bins or labeled drawers for storing small components like resistors, capacitors, connectors, and fuses for easy access.

c) Testing Area

- A dedicated space with tables or benches for testing and troubleshooting.
- Equipment includes multimeters, oscilloscopes, signal generators, power supplies, and circuit analysers.
- Test setups might involve secured mounting areas for circuit boards and devices.

d) Safety Features

- Clearly marked safety zones with fire extinguishers, first-aid kits, and eyewash stations.
- Proper ventilation for areas involving soldering or other heat-generating tasks.
- Anti-static mats and grounding points to protect sensitive electronics.

e) Storage for Materials

- Racks or closets for bulk storage of cables, wire spools, and larger equipment.
- Lockable cabinets for hazardous materials like chemicals or batteries.

f) Cable Management and Power Supply

- Organized cable management along the walls or workbenches.
- Easily accessible power outlets and circuit breakers, often mounted above workbenches or on walls.

18.1.2 Electrical Store Layout

1 Entrance and Showroom

- A welcoming entrance with clearly marked aisles for browsing.
- Display shelves showing featured items like electrical tools, home fixtures, and circuit kits.
- Bins or racks for promotional or discounted items.

2 Product Display

- Wall-mounted pegboards for displaying smaller packaged items like cables, plugs, switches, and adapters.
- Shelves with components like light bulbs, circuit breakers, and electrical panels.
- Display counters with locked glass cases for high-value items like power tools or specialized electronics.

3 Storage for Inventory

- Stockroom or back area with extra inventory, organized on shelves or in bins.
- Bulk storage for items like large rolls of cable or boxed tools.

4 Checkout Counter

- A centrally located or front-facing counter for customer assistance and purchases.
- Often equipped with a computer for inventory management, cash register, and a space for small packaging.

5 Special Sections

- Areas designated for lighting displays, demonstrating different light fixtures and bulbs.
- A section for DIY or hobby electronics with components like breadboards, microcontrollers, and basic circuit kits.

6 Signage and Navigation

- Clear signage indicating different product sections, such as "Cables & Wires," "Tools," "Lighting," etc.
- Aisles labelled and wide enough for easy navigation, accommodating carts or customer flow.

Both spaces prioritize a clean and organized layout to facilitate quick access to tools, components, and customer service. Safety considerations, efficient storage solutions, and a focus on accessibility are key aspects of a well-designed electrical workshop and store. An example is as shown in Figure 18.8.

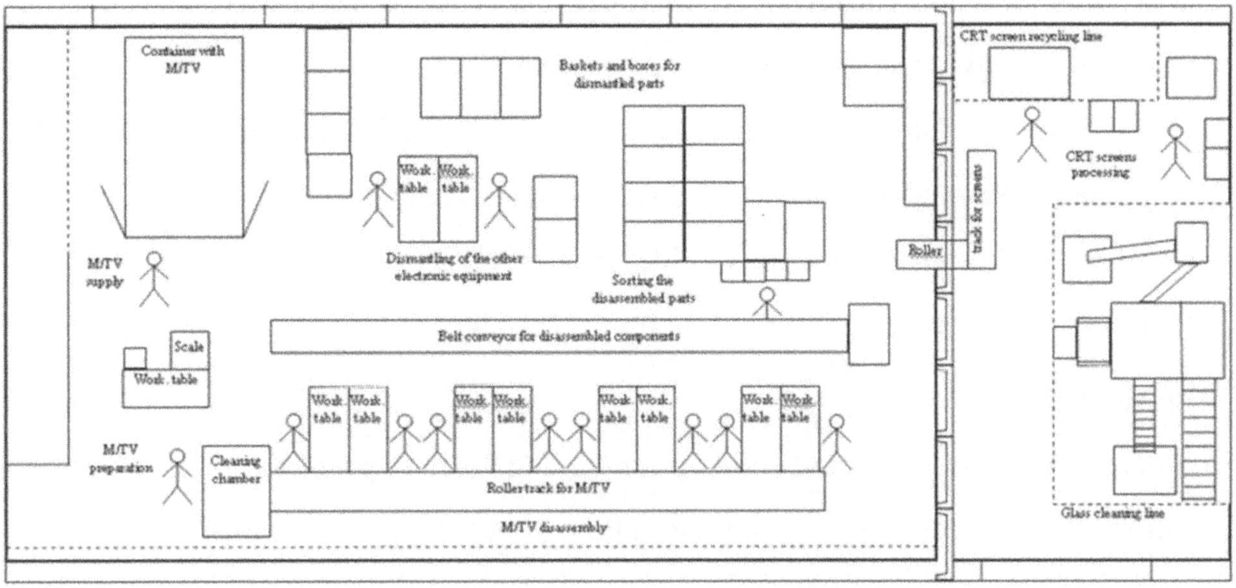

Figure 18.8: Example of Electrical Workshop Layout

18.2 TYPICAL INSTRUMENTS AND EQUIPMENT FOUND IN STORES

In an electrical workshop store, a variety of instruments and equipment cater to both professionals and hobbyists who work with electrical components.

18.2.1 List of the Most Common Items

1 Measuring and Testing Instruments

- **Multimeters** – For measuring voltage, current, and resistance.
- **Oscilloscopes** – For viewing and analysing waveforms in electronic circuits.
- **Clamp Meters** – To measure current without direct contact.
- **Insulation Testers** – To check the quality of insulation in electrical components.
- **Power Supplies** – Adjustable power sources for testing circuits.
- **Continuity Testers** – Simple tools to test if an electrical connection is intact.
- **Signal Generators** – Devices for generating electrical signals to test circuits.
- **Infrared Thermometers** – For non-contact temperature measurements.

2 Tools for Electrical Work

- **Screwdrivers** – Insulated types for working with electrical systems.
- **Pliers** – Including needle-nose, cutting, and crimping pliers.

- **Wire Strippers and Cutters** – For removing insulation and cutting wires.
- **Crimping Tools** – To attach connectors to cables.
- **Cable Testers** – For checking the functionality of network and coaxial cables.
- **Soldering Irons and Stations** – For assembling and repairing electronic components.
- **Desoldering Pumps** – To remove solder from joints.
- **Heat Guns** – For shrink-wrapping cables or desoldering.

3 **Power Tools**

- **Electric Drills** – Corded and cordless for drilling holes in panels or walls.
- **Rotary Tools** – For precise cutting, grinding, or polishing.
- **Angle Grinders** – For cutting metal or cleaning metal surfaces.
- **Dremel Tools** – For small-scale work on electrical components.

4 **Storage and Organization**

- **Component Organizers** – Small bins or cabinets for storing resistors, capacitors, connectors, and other small parts.
- **Tool Boxes and Cases** – For storing and transporting tools.
- **Spool Racks** – For organizing and dispensing different types of cables or wires.
- **Pegboards** – For hanging and displaying tools.

5 **Safety Equipment**

- **Insulated Gloves** – To protect hands when handling live circuits.
- **Safety Glasses** – For eye protection while cutting or soldering.
- **Anti-Static Mats** – To prevent static discharge when working with sensitive electronics.
- **Fire Extinguishers** – Specifically for electrical fires.
- **Ear Protection** – If power tools are frequently used in the workshop.

6 **Electrical Components and Supplies**

- **Cables and Wires** – Different gauges and types, including power cables, network cables, and coaxial cables.
- **Connectors** – Like terminal blocks, spade connectors, and plug connectors.

- **Fuses and Circuit Breakers** – Protection devices for circuits.
- **Relays and Switches** – For controlling electrical currents.
- **Light Fixtures** – Including LED, fluorescent, and incandescent lights.
- **Batteries** – Rechargeable and disposable in various sizes.
- **Power Outlets and Plugs** – Standard and specialized types.

7 **Soldering and Assembly Supplies**

- **Solder** – Wire or paste used for assembling electrical circuits.
- **Flux** – For cleaning and prepping metal surfaces before soldering.
- **Heat Shrink Tubing:** For insulating and bundling cables.
- **Electrical Tape** – Insulating tape for electrical work.
- **Breadboards** – For prototyping circuits without soldering.

8 **Kits and Spccialized Tools**

- **Arduino and Raspberry Pi Kits** – For prototyping and educational projects.
- **Circuit Board Kits** – For hobbyists and small-scale circuit projects.
- **Crimping Kits** – For terminating cables.
- **Tool Sets** – Complete sets for specific tasks like networking or automotive electrical work.

9 **Lighting and Home Electrical Supplies**

- **Light Bulbs** – Various types like incandescent, LED, and halogen.
- **Ballasts** – For fluorescent lighting systems.
- **Switch Panels** – Wall switches, dimmers, and smart switches.
- **Electrical Boxes and Conduit** – For housing and protecting electrical connections.

10 **Installation and Maintenance Tools**

- **Voltage Detectors** – For identifying live wires.
- **Fish Tapes** – For pulling cables through conduits or walls.
- **Conduit Benders** – For bending electrical conduits.
- **Stud Finders** – To locate studs or hidden wiring in walls.

18.2.2 Conclusion

This diverse range of tools and equipment helps ensure that both professional electricians and DIY enthusiasts have the right resources for electrical installation, maintenance, and repair tasks.

18.3 TESTING AND CALIBRATING ELECTRICAL MEASURING INSTRUMENTS

Testing and calibrating electrical measuring instruments are essential processes to ensure accuracy, reliability, and safety when working with electrical systems. Properly calibrated instruments deliver precise measurements, reduce errors, and help maintain high standards in electrical work.

18.3.1 Comprehensive Guide to the Testing and Calibration of Electrical Measuring Instruments

1 **Importance of Testing and Calibration**

- **Accuracy** – Calibrating electrical instruments ensures they provide accurate readings, reducing the risk of errors that could lead to faulty installations, incorrect diagnoses, or unsafe situations.
- **Safety** – Accurate instruments are crucial for safety when dealing with high-voltage systems, preventing incorrect readings that could cause electric shock or damage.
- **Compliance** – Many industries have strict regulations and standards for electrical measurements. Calibration helps ensure compliance with standards like ISO, IEC, or other regulatory guidelines.
- **Efficiency** – Well-calibrated tools save time and resources by providing reliable measurements, avoiding rework, and minimizing downtime.

2 **Types of Electrical Measuring Instruments**

- **Multimeters** – Measure voltage, current, and resistance.
- **Clamp Meters** – Measure current without direct contact.
- **Oscilloscopes** – Display waveforms for voltage and signal analysis.
- **Power Meters** – Measure power consumption in electrical systems.
- **Insulation Testers** – Check insulation resistance.
- **Signal Generators** – Create electrical signals for testing purposes.
- **Earth Ground Testers** – Measure grounding effectiveness.

- **Phase Rotation Testers** – Determine the phase sequence in three-phase systems.

3 Calibration Procedure

Calibration involves comparing an instrument's readings with a reference standard that has a known and accurate value. The steps are as follows:

a) Pre-Calibration Check

- Inspect the instrument visually for any physical damage, loose connections, or wear and tear.
- Ensure the instrument is clean, especially the connectors and probes, to avoid any influence on readings.

b) Reference Standards

- Use a high-precision reference instrument or calibration standard (also called a "master instrument") that is traceable to a national or international standard.
- These reference standards are usually maintained in controlled environments to ensure stability and accuracy.

c) Calibration Environment

- Perform calibration in a stable environment with minimal electrical interference, controlled temperature, and humidity.
- Environmental factors can affect the readings, so consistency is key.

d) Testing Procedure

- Connect the instrument under test to the reference standard using proper cables and connectors.
- Apply a known voltage, current, resistance, or signal from the standard to the instrument.
- Compare the readings of the instrument with the known values from the standard.
- Note any discrepancies between the instrument's readings and the reference values.

e) Adjusting the Instrument

- If discrepancies exceed acceptable tolerance levels, adjust the instrument's settings (if possible) to bring readings back to within the standard.
- Some instruments have internal adjustment mechanisms, while others may require software-based calibration adjustments.

f) Post-Calibration Documentation

- Record the results, adjustments made, and the instrument's final readings in a calibration certificate or report.
- Include details such as the calibration date, environmental conditions, and the reference standards used.

4 Testing Specific Instruments

- **Multimeters** – Use known standard resistors, voltage sources, and current sources. Compare the multimeter's readings with the reference.
- **Clamp Meters** – Pass a known current through a conductor and use the clamp meter to measure it. Compare the result with the standard.
- **Oscilloscopes** – Use a calibrated signal generator to create known waveforms. Check if the oscilloscope accurately displays the frequency, amplitude, and waveform shape.
- **Insulation Testers** – Measure the resistance of a known high-resistance reference. Compare the displayed insulation resistance with the standard.

5 Frequency of Calibration

- **Routine Calibration** – Typically performed annually, but high-usage instruments might require calibration more frequently.
- **After Repairs** – Instruments that have been repaired or altered should be recalibrated before use.
- **In Cases of Suspected Inaccuracy** – If there are reasons to believe an instrument is delivering incorrect readings, immediate calibration is necessary.
- **High-Precision Applications** – Instruments used in critical or high-precision applications may require more frequent calibration intervals.

6 Challenges in Calibration

- **Environmental Influence** – Temperature, humidity, and electrical interference can affect calibration accuracy.
- **Component Aging** – Over time, internal components like resistors and capacitors can drift, influencing accuracy.
- **Equipment Quality** – The quality and stability of the reference standards used directly influence the accuracy of the calibration.

7 Calibration Labs and Standards

- **In-House Calibration** – Some organizations have dedicated labs to calibrate instruments using high-precision equipment.
- **Third-Party Calibration** – External accredited calibration labs perform independent and certified calibration services.
- **Traceability** – All calibration should be traceable to national or international standards, ensuring consistent and globally recognized accuracy.

8 Record-Keeping and Calibration Labels

- **Calibration Records** – Keep detailed records of calibration results, adjustments, and any issues noted during the process. This helps track the instrument's performance over time.
- **Calibration Labels** – Place visible labels or stickers on calibrated instruments, indicating the calibration date, the due date for the next calibration, and the person or organization responsible.

18.3.2 Summary

In summary, regular testing and calibration of electrical measuring instruments are crucial to ensuring they deliver precise and reliable readings. Calibration helps avoid costly mistakes, maintains safety standards, and ensures compliance with industry regulations. It's a necessary practice for anyone involved in the installation, maintenance, and testing of electrical systems.

18.4 REGULAR MAINTENANCE AND SERVICING OF GENERATORS AND REWINDING MOTORS

Regular maintenance and servicing of generators and rewinding motors are essential for ensuring their longevity, reliability, and efficiency. A well-defined schedule helps prevent unexpected breakdowns, reduces downtime, and ensures the equipment operates at optimal performance.

18.4.1 Maintenance Schedule for Generators

1 Daily Maintenance (Pre-Operation Check)

- **Inspect Fuel System**
  - Check fuel levels and refill if necessary.
  - Inspect fuel lines and connections for leaks or blockages.

- **Check Oil Level**
  - Ensure the oil level is within the recommended range.
  - Look for signs of oil leaks around the engine.
- **Check Battery**
  - Test battery charge and condition.
  - Ensure the battery terminals are clean and tight.
- **Inspect Coolant System**
  - Verify coolant levels and look for signs of leaks in the radiator or hoses.
- **Examine Air Filters**
  - Check air filter for debris, dust, or clogging.
- **Visual Inspection**
  - Look for loose bolts, nuts, or parts.
  - Ensure there are no signs of corrosion or damage.

2 Weekly Maintenance

- **Clean the Generator**
  - Clean exterior surfaces to remove dust, dirt, and debris, especially around the vents and cooling fins.
- **Test Load**
  - Run the generator under load for a short period to ensure it performs under normal operating conditions.
- **Check Exhaust System**
  - Inspect the exhaust system for signs of blockages, leaks, or damage.
- **Check Fuel System**
  - Drain any water from the fuel tank or filter if applicable.
  - Inspect fuel injectors or pumps for any signs of wear or leakage.
- **Check the Battery Charger**
  - Test the battery charging system to ensure proper charging.

3 Monthly Maintenance

- **Oil Change**
  - Change the engine oil and replace the oil filter as per manufacturer recommendations.
- **Inspect and Clean Air Filter**
  - Clean or replace the air filter to ensure proper airflow.
- **Check Fuel Filter**
  - Inspect and replace the fuel filter if necessary.
- **Check and Test Safety Features**
  - Test the emergency stop system, low-oil pressure alarm, and high-temperature shutdown systems.
- **Inspect Belts and Hoses**
  - Check all belts, hoses, and connections for wear or damage. Replace as needed.

4 Quarterly Maintenance

- **Check the Generator's Alternator**
  - Inspect the alternator for wear, check brushes, and clean the alternator.
- **Check for Leaks**
  - Inspect the cooling system, oil system, and fuel system for leaks.
- **Lubricate Components**
  - Lubricate moving parts like bearings, linkages, and other mechanical components.
- **Examine Control Panel**
  - Ensure all meters, gauges, and indicators are working properly.
- **Inspect Electrical System**
  - Check wiring for signs of wear, corrosion, or damage.

5 Annual Maintenance

- **Complete Overhaul**
  - Perform a thorough inspection of all major components, including engine, alternator, control panel, and fuel system.
  - Check for carbon buildup and clean the combustion chamber.

- **Coolant and Radiator Check**
  - Flush and replace coolant as necessary.
  - Clean the radiator and check for signs of wear.
- **Check Engine Mounts**
  - Inspect and replace any worn or damaged engine mounts.
- **Battery Maintenance**
  - Perform a deep battery load test and replace the battery if it shows signs of deterioration.

18.4.2 Maintenance Schedule for Rewinding Motors

1 Daily Maintenance (Pre-Operation Check)

- **Check Motor Alignment**
  - Ensure the motor is properly aligned with the shaft or pulley.
- **Inspect for Overheating**
  - Check motor for any signs of overheating, like burning smells or excessive heat from the housing.
- **Lubrication Check**
  - Verify proper lubrication in bearings and other moving parts.
- **Inspect Power Supply**
  - Check incoming power supply voltage and compare it with motor requirements.

2 Weekly Maintenance

- **Clean the Motor**
  - Clean the motor casing and vents to remove dust, dirt, and debris that could affect airflow and cooling.
- **Check Electrical Connections**
  - Tighten any loose connections, inspect for signs of wear, corrosion, or overheating.
- **Check Insulation Resistance**
  - Test the insulation resistance of the motor windings using an insulation resistance tester (megger).

- **Check for Vibration**
  - Inspect for abnormal vibrations or noise, which could indicate misalignment or damage.

3 **Monthly Maintenance**

- **Inspect Bearings**
  - Check the motor bearings for wear and lubricate them.
- **Check and Clean Filters (If applicable)**
  - If the motor has filters or ventilation systems, clean or replace them as necessary.
- **Check Cooling Fans**
  - Inspect cooling fans for any blockage or dirt that may hinder airflow.
- **Inspect Motor Shaft**
  - Ensure the motor shaft is free of any rust or damage.

4 **Quarterly Maintenance**

- **Inspect Windings**
  - Perform a visual inspection of the windings for damage or wear. For larger motors, use a thermal imaging camera to detect hotspots.
- **Motor Current Test**
  - Measure and record the motor's amperage and compare it with the manufacturer's specifications.
- **Check Motor Starter and Control System**
  - Test the motor starter and control system to ensure proper operation.

5 **Semi-Annual Maintenance**

- **Rewind Motor (If Needed)**
  - If there is significant wear on the motor windings, schedule the rewinding process. This includes removing the old windings, cleaning the core, and installing new windings.
- **Detailed Inspection of Electrical Components**
  - Inspect all electrical components (contactors, relays, transformers) for signs of wear or deterioration.

6 Annual Maintenance

- **Comprehensive Overhaul**
  - Disassemble the motor for a complete internal inspection, including checking for signs of overheating, bearing wear, or insulation damage.
- **Test and Replace Worn Parts**
  - Replace worn bearings, brushes (for DC motors), or any damaged components.
- **Recalibrate Control Systems**
  - Check and recalibrate motor controllers, relays, and circuit breakers to ensure proper operation.
- **Re-inspect Windings**
  - Check the windings for degradation and perform an insulation resistance test.
- **Test Performance**
  - Run the motor under load to assess overall performance and efficiency, including checking for noise, vibration, or overheating.

18.4.3 General Tips for Both Equipment

- **Use Manufacturer's Guidelines** – Always refer to the manufacturer's manual for specific maintenance intervals and procedures.
- **Record Keeping** – Maintain accurate records of all maintenance performed, including inspections, repairs, and part replacements. This helps track the condition and longevity of equipment.
- **Professional Inspections** – For complex tasks like rewinding motors, it is often best to consult a professional motor technician or electrical engineer to perform advanced diagnostic checks.

18.4.4 Summary

In summary, by adhering to these schedules, both generators and rewinding motors can be kept in optimal condition, reducing the likelihood of unexpected failures and extending their operational lifespan.

18.5 EXAMINATION AND TESTING OF INSTALLED MOTORS AND SWITCHGEAR

The examination and testing of installed motors and switchgear are crucial for ensuring that electrical systems operate efficiently and safely. These components play a key role in industrial and commercial applications, and their reliability is paramount to avoid downtime, costly repairs, and safety hazards.

18.5.1 Examination and Testing of Installed Motors

a) Pre-Testing Procedures

Before starting any testing, ensure the following:

- **Power Supply** – Verify that the motor is connected to the correct power supply and that the supply voltage is within the motor's specified range.
- **Safety Protocols** – Confirm that all necessary safety precautions are in place, including isolating the motor from the power source and ensuring proper grounding.
- **Visual Inspection** – Inspect the motor for any visible signs of wear, overheating, or damage to components such as cables, connections, and the motor casing. Look for:
    - Burn marks, cracks, or corrosion.
    - Loose or disconnected wires.
    - Fluid leaks (from bearings or seals).
    - Dirty or blocked air vents.

b) Insulation Resistance Testing

- **Purpose** – To ensure the motor windings are properly insulated, preventing short circuits or electrical shocks.
- **Method** – Use an insulation resistance tester (megger) to measure the resistance of the windings to ground. The resistance value should typically be over 1 MΩ for motors, but this can vary depending on the motor's voltage rating and the manufacturer's specifications.
- **Steps**

i. Disconnect the motor from the power supply.

ii. Attach the insulation tester's leads to the motor windings and ground.

iii. Apply a voltage (usually 500V or 1000V depending on motor voltage).

iv. Read the resistance value; if the value is too low, the windings are defective and need repair or replacement.

c) Continuity Testing

- **Purpose**: To check the integrity of the motor windings.
- **Method**: Use a digital multimeter (DMM) to check for continuity in the motor windings.
- **Steps**

i. Set the multimeter to the continuity or resistance mode.

ii. Check the motor's windings by measuring between the terminals.

iii. A reading of zero or near-zero resistance indicates continuity. Infinite resistance (or no continuity) indicates an open winding or short circuit.

d) Rotor and Stator Test

- **Purpose**: To test for faults like rotor bars or stator winding short circuits.
- **Method**
  - **Rotating the Motor**: Spin the motor by hand (if possible) or run it under no load. Listen for any unusual sounds or vibrations that might indicate mechanical problems.
  - **Stator Current Testing**: If applicable, measure the current drawn by each phase of the stator under normal operating conditions. Imbalances could indicate faulty windings or rotor faults.

e) Vibration Analysis

- **Purpose**: To detect mechanical imbalances, misalignments, or bearing wear that could lead to motor failure.
- **Method**
  - Use a vibration meter or analyser to check the vibration levels of the motor during operation.
  - Compare the readings to normal vibration thresholds. Excessive vibration may point to problems such as loose mounting, misalignment, or faulty bearings.

f) Motor Performance Testing

- **Purpose**: To verify that the motor is performing to its specifications under load.
- **Method**

- Check the motor's current draw and compare it to the nameplate data.
- Measure the motor's speed (RPM) under load and compare it to the expected RPM.
- Check the motor is operating temperature (using a thermal camera or infrared thermometer). Excessive heat can indicate issues such as overloading, inadequate ventilation, or faulty windings.

g) Lubrication and Bearing Check

- **Purpose**: To ensure that the motor bearings are properly lubricated and functioning.
- **Method**
  - Check the oil level and condition of lubricants in the motor bearings.
  - For motors with grease-lubricated bearings, ensure the grease is fresh, and there is no evidence of over-greasing or under-greasing.
  - Inspect bearings for wear or play by rotating the shaft by hand and listening for any noise.

h) End of Testing

- After all tests, restore power and observe the motor running under load. Monitor its behavior for unusual sounds, excessive vibration, or overheating.
- Document all findings, including any adjustments, repairs, or replacements that are required.

18.5.2 Examination and Testing of Switchgear

Switchgear is essential for protecting electrical circuits, controlling the flow of electricity, and isolating faults in the system. Regular testing and maintenance of switchgear ensure operational reliability and safety.

a) Visual Inspection

- **Inspect the Switchgear Housing** – Look for any physical damage, corrosion, or signs of overheating in the switchgear panels, doors, and electrical connections.
- **Check for Proper Grounding** – Ensure that all switchgear components are properly grounded.
- **Examine Busbars and Connections** – Ensure all connections are tight, with no signs of arcing or excessive heating.
- **Inspect Control and Auxiliary Circuits** – Check relays, fuses, and control wires for wear or damage.

b) Insulation Resistance Testing

- **Purpose** – To verify that the insulation of the switchgear's busbars, cables, and components is intact.
- **Method** – Use an insulation resistance tester (megger) to test the busbars, cables, and connections to ground.
- **Steps**

1 Disconnect the switchgear from the power supply.

2 Apply the insulation tester to the conductors and the ground.

3 Record the resistance value; the value should typically be over 1 MΩ or as specified by the manufacturer.

c) Circuit Breaker Testing

- **Purpose** – To ensure that the circuit breakers function properly, protecting the system from overloads and faults.
- **Method**
  - **Operating Mechanism Check** – Test the operation of the breakers by manually tripping and resetting them.
  - **Test Trip Time** – Check the breaker's trip time to ensure it reacts quickly under fault conditions.
  - **Contact Resistance Test** – Measure the resistance of the breaker's contacts using a micro-ohmmeter. High resistance could indicate poor contact or arcing.
  - **Overcurrent Trip Test** – Simulate fault conditions to test the overcurrent protection mechanism of the breakers.

d) Protective Relay Testing

- **Purpose** – To verify that the protection relays are working correctly and responding to abnormal conditions like overvoltage, undervoltage, or overcurrent.
- **Method**
  - Test the relay's response to simulated fault conditions (e.g., overcurrent, earth fault).
  - Compare the relay's trip settings with the manufacturer's specifications.
  - Check the settings of time delay, trip current, and coordination with other protection devices.

e) Functional Testing of Switchgear Operations

- **Switch Operations** – Manually or remotely operate all switches, breakers, and isolators in the system to ensure smooth operation.
- **Interlocks and Safety Devices** – Verify that all interlocks and safety mechanisms are working correctly and preventing unsafe operations.
- **Functional Tests for Disconnectors** – Ensure that isolating switches (disconnectors) and earthing switches operate correctly.

f) Cable Testing

- **Purpose** – To verify that the cables connected to the switchgear are intact and in good condition.
- **Method** – Perform a continuity test and insulation resistance test on all cables to ensure they are free from defects.

g) End of Testing

- After completing all necessary tests, restore power to the system and operate the switchgear under load conditions.
- Monitor the system to ensure that the switchgear operates smoothly and reliably.

18.5.3 General Best Practices for Motor and Switchgear Testing

- **Record Keeping** – Maintain a detailed record of all tests, inspections, and maintenance activities for both motors and switchgear. This helps identify trends and potential future problems.
- **Routine Schedule** – Perform regular maintenance checks according to the manufacturer's recommendations and regulatory standards.
- **Calibration of Instruments** – Ensure that testing instruments (e.g., insulation testers, multimeters) are regularly calibrated to ensure accuracy.

18.5.4 Summary

In summary, by following these guidelines, you can ensure that both motors and switchgear are functioning optimally and safely, reducing the risk of failures and improving the reliability of your electrical systems.

18.6 LOCATING FAULTS IN MOTORS AND SWITCHGEAR

Faults in motors and switchgear can lead to equipment failure, safety hazards, and operational downtime. Understanding how to locate and diagnose these faults is crucial for effective maintenance and repair. Here is a guide on how to locate faults in motors and switchgear, based on common fault types and diagnostic techniques.

18.6.1 Faults in Motors

Motors can experience a variety of faults, including electrical, mechanical, and thermal issues. Below are steps to locate and diagnose some of the most common faults:

a) Common Faults in Motors

i. **Electrical Faults**

- **Short Circuits**: A short circuit in the windings can cause the motor to overheat, draw excessive current, or trip circuit breakers.
- **Ground Faults**: Occurs when the motor windings or cables make contact with the ground, leading to an increase in leakage currents.
- **Open Circuits**: Broken connections in the windings or power supply can cause the motor to stop functioning.
- **Imbalanced Voltage**: Uneven supply voltage can cause poor motor performance or lead to overheating.

ii. **Mechanical Faults**

- **Bearing Failure**: Worn-out bearings can cause noise, vibration, and eventually motor failure.
- **Rotor Faults**: Broken rotor bars or misalignment can affect motor performance and efficiency.
- **Overheating**: Caused by excessive current draw, poor ventilation, or defective windings.

iii. **Thermal Faults**

- **Overheating Due to Overload**: Caused by running the motor beyond its rated load.
- **Poor Ventilation**: If the motor's cooling system is blocked or inadequate, it can overheat.

b) Diagnostic Techniques for Locating Motor Faults

i. **Visual Inspection**

- Look for physical damage like broken or worn-out parts, burned windings, or excessive dirt and dust around the motor.
- Check for abnormal signs such as excessive heat, oil leaks, or smoke.

ii. **Insulation Resistance Testing**

- Use a **megger** (insulation resistance tester) to check the motor windings. A low resistance reading indicates that there is a problem with the insulation (e.g., short circuit or grounding).
- **Test Procedure**: Disconnect the motor from the power supply, connect the megger to the windings and the ground, and apply the test voltage. If the resistance is less than the manufacturer's specification (usually 1 MΩ or higher), it indicates a fault in the motor windings.

iii. **Current Measurement**

- Use a **clamp meter** to measure the current drawn by the motor. Compare this with the motor's rated current. A higher current draw could indicate overload, a short circuit, or rotor issues.
- **Imbalance in Current**: Measure the current in all three phases (for three-phase motors). Any imbalance can indicate issues like an open winding, faulty connections, or an imbalanced load.

iv. **Vibration Analysis**

- Excessive vibration may indicate mechanical faults such as bearing wear, misalignment, or unbalanced rotors. Use a **vibration meter** to detect abnormal vibrations and analyse the frequency to identify the specific cause.

v. **Thermographic (Infrared) Inspection**

- Use an **infrared camera** to detect hot spots in the motor. Excessive heat can indicate problems like overloaded windings, poor connections, or failing bearings.

vi. **Winding Resistance Testing**

- Use a **winding resistance tester** to check for imbalances in the motor windings. Significant differences in resistance between phases can indicate problems like open circuits or shorted turns in the windings.

vii. **Rotor Movement Test**

- Rotate the motor shaft by hand and listen for any unusual noises (grinding, squeaking) or feel for resistance (sign of bearing failure or misalignment).

viii. **Speed and Load Testing**

- Compare the actual speed of the motor (RPM) under load with the rated speed. A significant difference may indicate a mechanical problem like rotor damage or misalignment.

18.6.2 Faults in Switchgear

Switchgear faults can be classified into several categories: electrical faults, mechanical failures, and protection issues. These faults affect the protection and control of electrical circuits.

a) Common Faults in Switchgear

1 Electrical Faults

- **Short Circuits** – Caused by a fault in the insulation, leading to arcing and potential damage to the switchgear components.
- **Overcurrent** – Occurs when the current exceeds the rated value, typically due to faults downstream or overload conditions.
- **Earth Faults** – Can occur when there is an unintentional connection between a live conductor and the ground, causing current leakage.
- **Voltage Imbalance** – When one phase of a three-phase system carries more or less voltage than the others, it can damage switchgear components.

2 Mechanical Faults

- **Worn or Faulty Switch Contacts** – Over time, the contacts may wear out due to arcing, leading to poor conductivity, excessive heat, or even contact failure.
- **Failure of Moving Parts** – Mechanical failure in circuit breakers, switches, or relays can result in non-operation or failure to trip during faults.
- **Corrosion or Contamination** – External environmental factors can lead to corrosion of components like busbars, isolators, or circuit breaker terminals.

3 Protection and Control Faults

- **Tripping Failure** – The switchgear may fail to trip during fault conditions due to a faulty relay or breaker mechanism.

- **Incorrect Relay Settings** – If protection relay settings are incorrectly configured, they may fail to protect against faults.

b) Diagnostic Techniques for Locating Switchgear Faults

1 Visual Inspection

- Inspect switchgear for visible damage, corrosion, discoloration, or signs of arcing around terminals and contacts.
- Look for excessive heat on breakers, busbars, or connections.

2 Contact Resistance Testing

- Use a **micro-ohmmeter** to measure the resistance of the contacts in the switchgear. High contact resistance can cause heating and result in equipment failure.

3 Insulation Resistance Testing

- Use a megger to check the insulation resistance of busbars, cables, and other switchgear components. A low resistance reading suggests deterioration or degradation of insulation.

4 Relay Testing

- Test the operation of protection relays to ensure they are calibrated correctly. Simulate fault conditions (e.g., overcurrent, earth fault) and check if the relay trips as expected.
- Verify the settings of each relay and confirm they match the required protection characteristics.

5 Breaker Operation Testing

- Manually operate circuit breakers to ensure they open and close smoothly. Faulty or stiff breakers may indicate mechanical issues.
- **Test Trip Mechanism** – Simulate fault conditions to ensure that the breaker trips when the current exceeds the set threshold.

6 Current and Voltage Measurements

- Measure the current in each phase of a circuit and check for imbalance. Imbalance may indicate issues with the system or switchgear.
- Measure the voltage levels to confirm that they are within the acceptable range and to check for voltage imbalance.

7 **Thermal Imaging (Infrared Inspection)**

- Use infrared thermography to identify hot spots in switchgear components, such as contacts, circuit breakers, and busbars. Excessive heat is usually a sign of high resistance connections or arcing.

8 **Mechanical Inspection**

- Inspect the mechanical components of the switchgear such as hinges, springs, and locks. These components should operate freely without resistance. Lubrication may be required for moving parts.

9 **Functional Testing of Interlocks**

- Test the operation of interlocks and safety devices in the switchgear. These interlocks prevent improper operations, like closing a breaker when it's in the wrong position or not properly isolated.

10 **Busbar Inspection**

- Inspect the busbars for any signs of overheating or discoloration. Overheating can occur due to poor connections, excessive current, or contamination.

18.6.3 General Best Practices for Locating Faults

- **Use Proper Test Equipment** – Always use the appropriate and calibrated test instruments for diagnostics (e.g., megger, clamp meter, micro-ohmmeter).
- **Consult the Manufacturer's Manual** – Refer to the motor and switchgear manuals for specific troubleshooting steps and acceptable fault limits.
- **Document Findings** – Keep detailed records of all tests performed, including readings, fault symptoms, and any corrective actions taken.
- **Routine Inspections** – Perform regular inspections and preventive maintenance to identify potential issues before they lead to faults.
- **Expert Assistance** – For complex faults (e.g., rewinding motors or severe switchgear failures), consider consulting experienced professionals to handle repairs and replacements.

18.6.4 Summary

By following these guidelines, you can efficiently locate faults in motors and switchgear, enabling timely repairs and ensuring the continued safe operation of electrical systems.

18.7 REPAIR OF FAULTS IN MOTORS AND SWITCHGEAR

Repairing faults in motors and switchgear involves several steps to identify the issue, locate the source of the fault, and apply corrective actions. Below is an outline of the common faults and repair procedures:

18.7.1 Motor Faults and Repair

a) Common Motor Faults

- **Electrical Faults** – Short circuits, open circuits, or ground faults in the windings.
- **Mechanical Faults** – Bearing failure, rotor imbalance, misalignment, or shaft defects.
- **Overheating** – Caused by overloads, poor ventilation, or defective windings.
- **Insulation Breakdown** – Caused by aging or environmental factors.

b) Fault Diagnosis

- **Visual Inspection** – Check for physical damage such as broken parts, signs of overheating, or wear on bearings.
- **Testing for Continuity** – Use a multimeter or megger to check for open circuits or short circuits in windings.
- **Vibration Analysis** – Identify any misalignment or imbalance in the motor.
- **Temperature Monitoring** – Check if the motor is overheating, which may indicate an overload or poor cooling.

c) Repair Procedures

- **Rewinding**: If the windings are burnt or damaged, the motor will need to be rewound.
- **Bearing Replacement** – Replace worn-out bearings to restore motor function and reduce noise.
- **Cleaning and Lubrication** – Clean dirt and debris from the motor and lubricate moving parts such as bearings.
- **Replacing Faulty Parts** – Replace damaged or worn-out components such as rotor or stator parts.
- **Re-aligning** – Ensure the motor is properly aligned to prevent further mechanical stress.

18.7.2 Switchgear Faults and Repair

a) Common Switchgear Faults

- **Tripping or Failure to Trip** – The circuit breaker trips unexpectedly or fails to trip when needed.
- **Overheating** – Caused by excessive current, poor maintenance, or insufficient cooling.
- **Contact Wear** – Damage to contacts or insulation that affects the switchgear's functionality.
- **Arcing** – Can occur when opening or closing the switchgear under load, causing damage to the **contacts, insulation, and internal components** of the switchgear.

b) Fault Diagnosis

- **Visual Inspection** – Look for signs of overheating, damaged insulation, or burnt contacts.
- **Testing for Continuity** – Use a multimeter or insulation tester to check for proper continuity and insulation.
- **Circuit Breaker Testing** – Use a primary injection test set to ensure the breaker operates correctly under fault conditions.
- **Arc Fault Detection** – Check for signs of arcing on the contacts or bushings.

c) Repair Procedures

- **Contact Replacement** – Replace worn or damaged contacts in the switchgear to restore functionality.
- **Cleaning** – Clean the switchgear components to remove dirt, dust, and debris that may cause malfunction.
- **Lubrication** – Lubricate moving parts to reduce wear and tear.
- **Re-calibration of Protection Settings** – Ensure the protection settings (overcurrent, earth fault) are correctly configured.
- **Insulation Replacement** – Replace any damaged insulation to prevent electrical faults.

d) Preventive Maintenance

Regular maintenance for motors and switchgear helps prevent faults from occurring. This includes:

- **Cleaning and Inspection** – Periodically clean and inspect motors and switchgear to identify early signs of damage.

- **Testing** – Regular electrical testing (e.g., insulation resistance, continuity testing) helps detect potential issues.
- **Lubrication** – Ensure proper lubrication of bearings and moving parts.
- **Monitoring** – Install temperature or vibration sensors to detect early signs of motor or switchgear failure.

18.7.3 Conclusion

By following these procedures, maintaining a regular maintenance schedule, motors and switchgear can operate efficiently, and safely, minimizing downtime and costly repairs.

CHAPTER NINETEEN

19.0 REFRIGERATION AND AIR CONDITIONING

19.1 MAINTENANCE/SERVICING SCHEDULES FOR REFRIGERATION PLANT

A refrigeration plant, whether used in industrial, commercial, or residential settings, requires regular maintenance to ensure its smooth operation, energy efficiency, and longevity. The maintenance schedule should include daily, weekly, monthly, quarterly, and yearly tasks, covering all major components such as compressors, evaporators, condensers, expansion valves, and piping systems.

19.1.1 Comprehensive Maintenance Schedule for a Refrigeration Plant

1 Daily Maintenance

Objective – Ensure the system is running efficiently and detect early signs of any faults.

- **Check Operating Pressures** – Verify that suction and discharge pressures are within recommended limits.
- **Inspect Temperature Readings** – Ensure that the evaporator and condenser temperatures are within the required range.
- **Monitor Power Consumption** – Track electrical consumption to detect any unusual spikes or drops.
- **Inspect System for Leaks** – Check for any refrigerant leaks or unusual vibrations in the system.
- **Check Refrigerant Levels** – Ensure the refrigerant charge is at an appropriate level.
- **Examine Airflow** – Ensure proper airflow around the evaporator and condenser.
- **Inspect Filters** – Check air filters for dirt accumulation and clean if necessary.
- **Monitor Compressor Operation** – Listen for any unusual noises and ensure smooth operation.

2 Weekly Maintenance

Objective – Perform more detailed checks and clean key components to maintain system performance.

- **Clean Condenser Coils –** Remove any dirt, dust, or debris that could obstruct airflow and reduce efficiency.
- **Clean Evaporator Coils –** Check the evaporator coils for any ice build-up or dirt and clean them as needed.
- **Check Compressor Oil –** Inspect oil levels in the compressor, and top up or change oil if necessary.
- **Inspect Piping for Leaks –** Check refrigerant lines for any signs of wear or leaks, especially at joints and connections.
- **Examine Vibration Levels –** Measure the vibration levels of the compressor, and check motor and pump alignment.
- **Check Drainage –** Ensure that drainage systems (for condensate) are not clogged and working properly.

3 Monthly Maintenance

Objective – Thorough inspection and testing of critical components for reliability.

- **Test Safety Controls –** Verify the operation of safety devices such as pressure switches, temperature cutoffs, and overload protection.
- **Inspect Expansion Valves –** Check for correct operation of the expansion valves, ensuring they are free from blockages.
- **Inspect and Clean Air Handling Units –** Clean air filters, coils, and drain pans to ensure good airflow and prevent clogging.
- **Inspect Electrical Connections –** Tighten and inspect electrical connections for wear, corrosion, or overheating.
- **Check Fan Belts –** Inspect and adjust fan belts for tension and wear.
- **Monitor Oil and Refrigerant Levels –** Perform a more detailed check of oil and refrigerant levels to ensure the plant is operating optimally.

4 Quarterly Maintenance

Objective – Conduct an in-depth analysis of system performance and identify potential issues before they escalate.

- **Inspect System Efficiency –** Perform an efficiency test to compare actual performance against design specifications.

- **Check Heat Exchanger –** Inspect and clean the heat exchangers to ensure they are not clogged or reducing system efficiency.
- **Test Emergency Shutdown Function –** Test the emergency shutdown system to ensure it operates as intended in case of system failure.
- **Inspect Insulation** – Check the insulation around pipes and equipment for damage or wear.
- **Check Refrigerant Pressure and Quality** – Test refrigerant pressure and quality to ensure the system is charged correctly.
- **Inspect Control Panel and Wiring** – Verify the functionality of the control panel and ensure wiring is secure and in good condition.

5 Annual Maintenance

Objective – Conduct a comprehensive check of all components to ensure the plant is in top condition for the coming year.

- **Comprehensive Cleaning** – Clean all components, including coils, filters, fans, and drains.
- **Full Refrigerant Charge Check** – Conduct a detailed refrigerant charge test and top off or recover refrigerant as needed.
- **Inspect and Test Compressor** – Perform a detailed inspection and test the compressor for wear, checking oil levels, valve conditions, and overall functionality.
- **Check Heat Recovery System (if applicable)** – Inspect any heat recovery systems, ensuring efficient operation.
- **Perform Performance Calibration** – Test and recalibrate the system's performance to ensure it is operating at optimal levels.
- **Check and Replace Air Filters** – Replace old air filters with new ones, ensuring air quality is maintained.
- **System Leak Check –** Perform a comprehensive leak test to ensure no refrigerant loss, including inspecting components that are difficult to access.

19.1.2 Additional Recommendations

- **Keep a Maintenance Log** – Document all maintenance activities, including tasks completed, parts replaced, and observations. This helps with tracking performance trends and scheduling future maintenance.

- **Monitor System Alarms and Alerts** – Set up monitoring systems with alarms to alert operators of unusual conditions (e.g., over-temperature, pressure issues).
- **Staff Training** – Ensure that maintenance personnel are adequately trained to handle emergency situations and repairs on the refrigeration plant.

19.1.3 Conclusion

By following this maintenance schedule, the lifespan of the refrigeration plant can be extended, ensuring energy efficiency, and avoiding costly repairs due to unforeseen breakdowns. Regular maintenance helps identify issues early, optimizing the plant's performance and reliability.

19.2 TROUBLESHOOTING TECHNIQUES FOR A REFRIGERATION PLANT

Troubleshooting a refrigeration plant involves systematically diagnosing and resolving issues to restore normal operation. Common issues in refrigeration plants include temperature fluctuations, unusual noises, system inefficiency, and compressor failure.

19.2.1 Compressor Problems: Symptoms, Causes, and Troubleshooting Steps

1 System Not Cooling Properly

a) Symptoms

- The refrigeration unit is not maintaining the desired temperature.
- The area or space being cooled is warmer than expected.

b) Possible Causes

- **Low Refrigerant Charge** – A drop in refrigerant levels can reduce the system's cooling efficiency.
- **Dirty Condenser Coils** – Dirt and debris can obstruct airflow, reducing heat exchange efficiency.
- **Clogged Expansion Valve** – The expansion valve may be blocked or malfunctioning, hindering refrigerant flow.
- **Faulty Thermostat** – Incorrect thermostat settings or a malfunctioning thermostat can cause the system to fail to cool properly.

c) Troubleshooting Steps

- **Check Refrigerant Levels** – Use a gauge to check the refrigerant pressure on the suction and discharge lines. If the pressure is too low, the system likely has a refrigerant leak.

- *Solution:* If low, recharge the system with the correct refrigerant and repair any leaks.

- **Inspect Condenser Coils** – Clean the condenser coils with a soft brush or coil cleaner to remove dirt and debris.
- **Check Expansion Valve** – Ensure the expansion valve is not blocked. If necessary, replace or clean it.
- **Test the Thermostat** – Use a thermometer to verify that the thermostat is operating correctly. Adjust or replace the thermostat as needed.

2 Compressor Not Starting

a) Symptoms

- The compressor does not start, even when the system is calling for cooling.
- The compressor motor hums but does not start.

b) Possible Causes

- **Electrical Issues** – A blown fuse, tripped breaker, or faulty wiring can prevent the compressor from starting.
- **Overload Protection** – The compressor's internal overload protection may have triggered due to overheating or electrical surges.
- **Faulty Start Capacitor** – A malfunctioning capacitor can prevent the compressor from starting properly.
- **Frozen Condenser** – A build-up of ice on the condenser can cause excessive pressure, triggering the overload protection.

c) Troubleshooting Steps

- **Check Electrical Connections** – Inspect fuses, circuit breakers, and wiring to ensure there are no electrical issues.
    - *Solution*: Reset or replace any tripped breakers, fuses, or faulty wires.
- **Test the Overload Relay** – If the compressor is not starting, check the overload relay and reset it or replace it if necessary.
- **Check the Start Capacitor** – Use a multimeter to test the start capacitor. Replace it if it shows signs of failure (e.g., no continuity).
- **Inspect the Condenser** – If the condenser is frozen, turn off the system to allow it to defrost. Check airflow and clean the condenser.

3 Excessive Vibration or Noise

a) Symptoms

- Unusual noises such as rattling, buzzing, or grinding coming from the refrigeration unit.
- Excessive vibrations when the system is running.

b) Possible Causes

- **Loose Components** – Loose parts such as compressor mounts, fans, or fan blades can cause vibrations and noise.
- **Worn Bearings** – Worn bearings in the compressor or fans can cause a grinding noise.
- **Misalignment**: Misaligned components (e.g., motor shaft or compressor) may generate vibration.

c) Troubleshooting Steps

- **Tighten Loose Components** – Inspect the unit for any loose bolts, screws, or panels, and tighten them.
- **Check Compressor Bearings** – Listen for grinding noises. If the compressor bearings are worn, they may need to be replaced.
- **Inspect Fan Blades** – Check for bent or damaged fan blades that could cause vibrations and noise.
- **Verify Alignment** – Ensure that all components (compressor, motor, fan) are properly aligned. Misalignment can be corrected by realigning the components or replacing worn parts.

4 High Pressure in the System

a) Symptoms

- The system is operating at higher-than-normal discharge pressures, leading to compressor strain or system shutdown.

b) Possible Causes

- **Blocked Condenser** – A blocked or dirty condenser coil reduces the heat dissipation process, causing high pressure.
- **Overcharged Refrigerant** – Excess refrigerant in the system increases pressure and could cause failure.

- **Inadequate Airflow** – Poor airflow due to dirty filters, fans, or coils may cause high pressure in the system.
- **Faulty Pressure Relief Valve** – If the pressure relief valve is defective, it may not regulate the pressure correctly.

c) **Troubleshooting Steps**

- **Check Refrigerant Charge** – Use gauges to check for overcharging. If the refrigerant level is too high, recover excess refrigerant to bring it to the optimal level.
- **Clean Condenser Coils** – Clean the condenser coils to remove any obstructions that could affect heat transfer.
- **Ensure Proper Airflow** – Clean air filters, and check that fans are working properly to ensure adequate airflow over the condenser.
- **Test the Pressure Relief Valve** – Ensure that the pressure relief valve is functioning correctly. If not, replace the valve.

5 **Frost or Ice Build-Up on Evaporator Coils**

a) **Symptoms**

- Ice or frost forms on the evaporator coils, restricting airflow and reducing cooling efficiency.

b) **Possible Causes**

- **Low Refrigerant Levels** – Low refrigerant can cause the evaporator coils to freeze up due to reduced heat absorption.
- **Airflow Issues** – Blocked or dirty air filters, fans, or coils can restrict airflow, causing moisture to freeze on the evaporator.
- **Faulty Defrost Timer** – In some systems, a malfunctioning defrost timer can prevent regular defrost cycles, leading to ice build-up.

c) **Troubleshooting Steps**

- **Check Refrigerant Charge** – Ensure the refrigerant levels are adequate. If low, recharge and check for leaks.
- **Inspect Air Filters** – Clean or replace any clogged air filters to ensure proper airflow.
- **Check Defrost Timer** – Inspect the defrost timer and replace it if it is not cycling properly.
- **Clean Evaporator Coils** – Remove any ice build-up from the coils and clean them to ensure proper heat exchange.

6 System Leaks

a) Symptoms

- Refrigerant leaks can result in poor cooling performance, high pressures, and system inefficiencies.

b) Possible Causes

- **Loose Connections** – Leaks can occur at joints, connections, or valves.
- **Damaged Piping** – Physical damage to refrigerant lines or fittings can cause refrigerant to leak.
- **Worn Seals or Gaskets** – Worn or damaged seals around components like the compressor or evaporator can cause leaks.

c) Troubleshooting Steps

- **Use a Leak Detector** – Employ a refrigerant leak detector or soapy water solution to find the source of the leak.
- **Check Joints and Connections** – Tighten any loose connections or replace any faulty seals or gaskets.
- **Inspect Refrigerant Lines** – Visually inspect the piping and components for damage. Replace any sections of piping that are compromised.
- **Recharge the System** – Once the leak is repaired, recharge the system with the correct refrigerant.

19.2.2 Summary

In summary, effective troubleshooting requires a methodical approach, attention to detail, and a sound understanding of how each component in the refrigeration system functions. By diagnosing and resolving faults quickly, refrigeration plants can operate efficiently, minimize downtime, and avoid costly repairs. Regular maintenance and periodic checks also help in preventing most common issues before they lead to system failure.

19.3 INSTALLATION PROCEDURES FOR AIR CONDITIONING EQUIPMENT

Installing air conditioning (AC) equipment requires a systematic and precise approach to ensure efficient and safe operation. Proper installation enhances the system's performance, longevity, and energy efficiency.

19.3.1 Step-By-Step Guide for Installing an Air Conditioning System

1 Pre-Installation Preparations

Objective: Ensure the installation site is suitable and prepare necessary tools and equipment.

- **Choose the Right Location** – Select an appropriate spot for both the indoor and outdoor units:
  - ***Indoor Unit*** – Install in a central location that allows even air distribution, away from heat sources (e.g., sunlight, appliances), and with enough clearance for air circulation.
  - ***Outdoor Unit*** – Place in a shaded, well-ventilated area with enough space for heat dissipation and easy access for maintenance.
- **Check Power Requirements** – Verify the power supply meets the voltage, current, and breaker specifications of the AC unit. Ensure proper grounding to prevent electrical hazards.
- **Gather Tools and Materials** – Prepare necessary tools such as a drill, pipe cutter, vacuum pump, refrigerant gauge, flare tools, spanners, and other equipment for electrical and refrigerant line work.

2 Install the Indoor Unit

Objective: Secure the indoor unit in the chosen location and make the necessary connections.

- **Mount the Wall Bracket** – If installing a wall-mounted split AC unit, fix the wall bracket securely using screws and a level. Ensure the bracket is level to prevent condensate drainage problems.
- **Create a Hole for the Piping** – Drill a hole (typically around 2.5–3 inches in diameter) through the wall to route the copper pipes, drain hose, and electrical cables to the outdoor unit.
  - ***Ensure Proper Angle*** – Tilt the hole slightly downward toward the outside to facilitate drainage.
- **Install Copper Pipes** – Attach the copper refrigerant lines to the indoor unit using flare nuts. Use a flare tool to prepare the ends if necessary.
  - ***Insulate Pipes*** – Wrap the pipes with insulation to prevent condensation and energy loss.
- **Connect the Drain Hose** – Attach the drain hose to the indoor unit, ensuring it slopes downward to allow water to flow out freely.
- **Connect the Electrical Wiring** – Follow the manufacturer's wiring diagram to connect the control cables to the indoor unit. Ensure proper terminals are used for connections.

3 Install the Outdoor Unit

Objective: Secure the outdoor unit in a suitable location and connect it to the indoor unit.

- **Place the Outdoor Unit** – Install the outdoor unit on a stable surface like a concrete pad or a metal frame. Ensure it is level to prevent vibration and noise.
  - *Mount on Brackets (Optional)* – If required, use anti-vibration pads or brackets to elevate the unit, especially in areas prone to flooding.
- **Connect Refrigerant Lines** – Use flare nuts to connect the copper refrigerant lines from the indoor unit to the outdoor unit.
  - *Avoid Bends and Kinks* – Keep the lines as straight as possible. Use a pipe bender if necessary to avoid sharp bends.
  - *Pressure Test* – Perform a pressure test to check for leaks in the refrigerant lines using nitrogen gas.
- **Electrical Connections** – Connect the electrical cables from the indoor unit to the outdoor unit, following the wiring diagram provided by the manufacturer.
- **Attach the Drain Hose** – If the drain hose from the indoor unit needs to be extended, secure it to a proper drainage area.

4 Vacuum the System and Check for Leaks

Objective: Remove moisture and air from the system and ensure the integrity of the refrigerant lines.

- **Vacuum the Refrigerant Lines** – Use a vacuum pump to evacuate air and moisture from the copper refrigerant lines. Connect the vacuum pump to the service port on the outdoor unit and run it until it reaches the recommended vacuum level (usually around 500 microns).
- **Check for Leaks** – Conduct a vacuum leak test to ensure there are no leaks in the lines.
  - *Hold Vacuum* – Turn off the pump and observe the vacuum gauge. If the pressure remains stable for 15–30 minutes, the system is leak-free.
- **Charge the System with Refrigerant** – Release the refrigerant from the outdoor unit's service valves into the system, as most new units come pre-charged with the correct refrigerant amount.

5 Test and Calibrate the System

Objective: Ensure the system operates correctly and efficiently after installation.

- **Power On the System** – Switch on the power supply and test the air conditioner by setting it to the desired temperature.
- **Check for Proper Functioning** – Observe if the unit cools properly, listen for any unusual noises, and check the air distribution from the indoor unit.
- **Calibrate the Thermostat** – Make adjustments to the thermostat to ensure the set temperature is maintained accurately.

Verify Refrigerant Pressure – Use a refrigerant gauge to check the pressure levels in the system. Ensure it aligns with the manufacturer's recommended pressure range.

- **Check Drainage** – Ensure the condensate is draining correctly from the indoor unit's drain hose without leaks.

6 Finishing Touches and User Guidance

Objective: Complete the installation and educate the user on the AC system's operation and maintenance.

- **Seal Wall Openings** – Use a sealant to close any gaps or holes created during the piping process to prevent air leaks and pests.
- **Secure Wires and Pipes** – Use cable ties or clips to secure electrical wires and pipes to avoid clutter.
- **User Orientation** – Explain to the user how to operate the unit, including how to use the remote control, change settings, and clean filters.
- **Schedule Maintenance** – Recommend a regular maintenance schedule, including cleaning filters, checking refrigerant levels, and inspecting electrical connections.

19.3.2 Common Installation Examples

1 Split Air Conditioning System

- **Installation of Indoor Unit** – Mounting the wall bracket, connecting refrigerant lines, and ensuring proper electrical connections.
- **Installation of Outdoor Unit** – Placing on a level surface, connecting refrigerant lines, and ensuring proper airflow.

2 Window Air Conditioner

- **Installation** – Fixing the unit in a window frame, sealing gaps, and plugging in the power source.

- **Consideration** – Ensuring the unit tilts slightly outward for proper water drainage.

3 **Central Air Conditioning System**

- **Installation** – Setting up ductwork, installing the central unit in the basement or utility room, and ensuring correct airflow through the house.
- **Consideration** – Ensuring all rooms have proper duct openings and registers for even cooling.

19.3.3 Summary

In summary, proper installation is crucial for the efficient and safe operation of air conditioning systems. Each step, from pre-installation preparation to final calibration, plays a vital role in ensuring optimal performance. A thorough understanding of the manufacturer's guidelines and adherence to safety standards is essential during installation. Regular maintenance after installation ensures the longevity of the system and prevents costly breakdowns.

19.4 MAINTENANCE/SERVICING SCHEDULES FOR AIR CONDITIONING PLANT

Regular maintenance of an air conditioning (AC) plant is crucial to ensure efficient operation, prolong equipment life, and minimize energy consumption.

19.4.1 Comprehensive Maintenance Schedules

Daily Maintenance Tasks

- **Visual Inspection**
  - Inspect the indoor and outdoor units for any visible signs of damage, leaks, or unusual noises.
  - Check for any obstructions around the outdoor unit to ensure proper airflow.
- **Monitor System Operation**
  - Check thermostat settings to ensure proper temperature control.
  - Observe any fluctuations in cooling performance or unusual vibrations.

Weekly Maintenance Tasks

- **Clean Air Filters**
  - Remove and clean air filters of indoor units to maintain efficient airflow and improve air quality.
  - Replace if necessary if they are damaged or heavily clogged.
- **Check Drainage System**
  - Inspect the condensate drain pan and drain line to ensure proper drainage and prevent clogging.
  - Clear any blockages to avoid water damage or mold growth.

Monthly Maintenance Tasks

- **Check Refrigerant Levels**
  - Use a gauge to measure refrigerant pressure levels. Ensure they are within manufacturer-recommended ranges.
  - Inspect refrigerant lines for any signs of leaks or damage.
- **Clean Coils**
  - Clean the evaporator and condenser coils to remove dust and debris that may hinder heat exchange efficiency.
  - Ensure the coil fins are straight and not damaged.
- **Lubricate Moving Parts**
  - Lubricate motors, fans, and bearings to reduce friction and wear.
  - Check the fan belt tension and adjust if necessary to prevent slippage.
- **Inspect Electrical Components**
  - Check all electrical connections, tighten loose terminals, and inspect wiring for wear or damage.
  - Test capacitors, relays, and contactors for proper operation.

Quarterly Maintenance Tasks

- **Check Thermostat Accuracy**
  - Calibrate the thermostat to ensure accurate temperature readings.
  - Replace batteries in programmable thermostats if needed.

- **Inspect Fan Blades**
  - Check the condition of fan blades in both indoor and outdoor units for damage, dirt buildup, or imbalance.
  - Clean blades as needed and tighten the fan blade mountings.
- **Test Safety Controls**
  - Inspect and test safety controls such as pressure switches, high-temperature cutoffs, and limit controls.
  - Ensure they are functioning correctly to prevent equipment damage.
- **Inspect Ductwork (if applicable)**
  - Inspect the duct system for leaks, damage, or debris buildup.
  - Seal any leaks using appropriate sealant materials.

Semi-Annual Maintenance Tasks (Every 6 Months)

- **Deep Cleaning of Indoor and Outdoor Units**
  - Perform a thorough cleaning of both indoor and outdoor units, including coil fins, fan blades, and casings.
  - Remove any accumulated debris around the outdoor unit for optimal airflow.
- **Check System Components for Wear**
  - Inspect belts, pulleys, bearings, and seals for signs of wear or damage.
  - Replace any worn-out parts to prevent potential breakdowns.
- **Test Refrigerant Leak Detection**
  - Use a refrigerant leak detector to check for small leaks in the system.
  - Repair leaks and recharge refrigerant if necessary.
- **Inspect Insulation**
  - Check insulation on refrigerant lines to ensure they are intact and in good condition.
  - Replace any damaged insulation to prevent energy loss.

Annual Maintenance Tasks

- **Inspect and Test Overall System Performance**
  - Conduct a comprehensive system check, including cooling efficiency, airflow, and electrical consumption.

- Analyse system performance data to identify potential issues or areas for improvement.
- **Replace Air Filters**
- Replace air filters even if they are not heavily clogged to ensure optimal air quality.
- Consider installing high-efficiency particulate air (HEPA) filters if needed.
- **Check and Clean Heat Exchangers**
- Clean heat exchangers thoroughly to ensure efficient heat transfer.
- Inspect for any signs of corrosion, scale buildup, or damage.
- **Test and Adjust Air Balancing**
- Adjust airflow settings to ensure even cooling across all areas served by the AC plant.
- Test the static pressure in the ducts and make necessary adjustments to dampers and diffusers.
- **Check Electrical System for Wear**
- Inspect electrical panels, control boards, and other critical electrical components.
- Test backup systems and ensure electrical connections are tight and free of corrosion.
- **Re-Calibrate Controls and Sensors**
- Re-calibrate temperature and humidity sensors to maintain accurate readings.
- Update control settings based on changes in occupancy or seasonal requirements.

19.4.1.1 Table of Example of Schedule Overview

Frequency	Tasks
Daily	Visual inspection, monitor system operation.
Weekly	Clean air filters, check drainage system.
Monthly	Check refrigerant levels, clean coils, lubricate parts, inspect electrical components.
Quarterly	Check thermostat accuracy, inspect fan blades, test safety controls, inspect ductwork.
Semi-Annual	Deep cleaning of units, check for wear, refrigerant leak detection, inspect insulation.
Annual	System performance test, replace air filters, clean heat exchangers, air balancing, electrical inspection.

19.4.2 Additional Maintenance Tips

- **Keep Maintenance Logs** – Maintain a detailed log of all maintenance activities, repairs, and adjustments for future reference. This helps track the AC plant's performance over time.
- **Follow Manufacturer Guidelines** – Always refer to the manufacturer's manual for specific maintenance recommendations, particularly regarding refrigerant types and pressures.
- **Use Professional Services** – Schedule annual professional inspcctions by certified IIVAC technicians to handle advanced diagnostics, refrigerant charging, and system calibration.

19.4.3 Summary

In summary, a well-maintained air conditioning plant operates efficiently, uses less energy, and has fewer unexpected breakdowns. Adhering to a structured maintenance schedule not only extends the lifespan of the equipment but also ensures a comfortable and safe environment for occupants.

19.5 IDENTIFICATION OF SYMPTOMS, LOCATION AND REPAIR OF FAULTS IN AIR CONDITIONING PLANT

Faults in air conditioning (AC) plants can significantly affect performance, efficiency, and comfort. Identifying symptoms, diagnosing the root cause, and performing repairs are crucial steps in maintaining an effective AC system.

19.5.1 Common Symptoms and Faults in Air Conditioning Plants

1 Symptom: Inadequate Cooling or No Cooling

- *Possible Causes*
  - Low refrigerant levels or leaks.
  - Dirty air filters or clogged evaporator/condenser coils.
  - Faulty compressor or compressor failure.
  - Blocked or leaking ducts.
  - Malfunctioning thermostat.
- *Diagnosis*
  - Check refrigerant pressure using a gauge to identify low levels.
  - Inspect air filters, evaporator coils, and condenser coils for dirt or blockages.

- Listen for unusual noises from the compressor that may indicate mechanical issues.
- Use a thermometer to compare the thermostat settings with the actual room temperature.

o ***Repair***

- Recharge refrigerant if low and repair any identified leaks.
- Clean or replace dirty air filters and coils.
- Repair or replace the compressor if damaged.
- Inspect and seal any leaks in the duct system.
- Recalibrate or replace a malfunctioning thermostat.

2 Symptom: AC Unit is Short Cycling (Turning On and Off Frequently)

o ***Possible Causes***

- Refrigerant overcharge or undercharge.
- Faulty thermostat or incorrect placement.
- Clogged air filter causing restricted airflow.
- Overheating compressor due to electrical issues or mechanical faults.

o ***Diagnosis***

- Check refrigerant levels for overcharge/undercharge.
- Inspect the thermostat for placement near heat sources or incorrect settings.
- Evaluate the condition of the air filter and measure airflow.
- Inspect electrical connections for short circuits and evaluate the compressor for overheating.

o ***Repair***

- Adjust refrigerant to correct levels.
- Relocate or recalibrate the thermostat if necessary.
- Clean or replace the air filter to improve airflow.
- Repair or replace the compressor if overheating or electrical issues are detected.

3 Symptom: Water Leaking from the Indoor Unit

o ***Possible Causes***

- Clogged or blocked condensate drain line.

- Rusted or cracked drain pan.
- Frozen evaporator coil due to restricted airflow or refrigerant issues.

o ***Diagnosis***

- Inspect the condensate drain line for blockages or clogs.
- Check the condition of the drain pan for rust or damage.
- Evaluate the evaporator coil for ice buildup, indicating poor airflow or low refrigerant.

o ***Repair***

- Clear the drain line using a vacuum or drain cleaner.
- Replace a damaged or rusted drain pan.
- Fix airflow issues by cleaning or replacing air filters, and check refrigerant levels for leaks.

4 Symptom: Unusual Noises from the AC Unit

o ***Possible Causes***

- Loose or worn-out fan belts.
- Failing or misaligned fan blades.
- Motor bearings worn out or lack lubrication.
- Debris or foreign objects in the unit.

o ***Diagnosis***

- Inspect fan belts for wear and proper tension.
- Check the alignment and condition of fan blades.
- Evaluate the condition of motor bearings for signs of wear or overheating.
- Examine the unit for debris or foreign objects obstructing airflow.

o ***Repair***

- Replace worn or loose fan belts.
- Realign or replace damaged fan blades.
- Lubricate or replace motor bearings if worn out.
- Remove debris and clean the unit thoroughly.

5 Symptom: Foul Odors Coming from the AC Unit

- ***Possible Causes***
  - Mold or mildew growth in the unit or ducts.
  - Clogged or dirty air filters.
  - Dead rodents or debris in the ductwork.
- ***Diagnosis***
  - Inspect the evaporator coil and drain pan for mold or mildew buildup.
  - Check air filters for cleanliness.
  - Inspect ductwork for dead animals or other debris.
- ***Repair***
  - Clean and disinfect the evaporator coil and drain pan with an anti-microbial solution.
  - Replace dirty air filters with new ones.
  - Remove debris and clean the ducts, sealing any openings to prevent pests.

6 Symptom: High Energy Consumption

- ***Possible Causes***
  - Dirty coils reducing heat exchange efficiency.
  - Low refrigerant levels causing the system to overwork.
  - Poor insulation in ducts leading to energy loss.
  - Faulty thermostat not reading temperatures accurately.
- ***Diagnosis***
  - Check the cleanliness of coils.
  - Measure refrigerant levels and inspect for leaks.
  - Inspect ductwork for insulation quality.
  - Test the thermostat's accuracy using a thermometer.
- ***Repair***
  - Clean the evaporator and condenser coils.
  - Recharge refrigerant to appropriate levels and fix leaks.
  - Re-insulate ducts with proper materials.

- Replace or recalibrate the thermostat if needed.

19.5.2 Common Tools for Troubleshooting AC Units

1 **Refrigerant Gauges** – For measuring refrigerant pressure levels.

2 **Digital Thermometer** – For checking air temperature at different points.

3 **Multimeter** – For electrical diagnostics like checking continuity, voltage, and current.

4 **Leak Detector** – For identifying refrigerant leaks.

5 **Vacuum Pump** – To clear clogged condensate lines.

6 **Coil Cleaner** – For cleaning dirty coils.

7 **Screwdrivers and Wrenches** – For disassembling parts and accessing internal components.

19.5.3 Best Practices for Fault Diagnosis and Repair

- **Systematic Troubleshooting** – Use a step-by-step approach to isolate the problem. Start with basic checks (like airflow and thermostat settings) before moving to complex issues (like refrigerant levels and compressor function).
- **Detailed Documentation** – Keep a record of symptoms, faults, and repairs for each unit. This helps in identifying recurring issues and aids in future maintenance.
- **Safety Precautions** – Always turn off power before working on electrical components. Use protective gear and follow safety guidelines to avoid injuries.
- **Follow Manufacturer Guidelines** – Use the manufacturer's service manual for troubleshooting specific models, as each may have unique diagnostic procedures.

19.5.4 Summary

In summary, regular troubleshooting and prompt repairs ensure that an air conditioning plant operates efficiently, minimizing downtime and energy consumption. Identifying symptoms early and addressing them prevents minor faults from escalating into major issues.

CHAPTER TWENTY

20.0 HEATING AND VENTILATION SYSTEMS

Heating and ventilation systems are essential components in residential, commercial, and industrial buildings. These systems rely on various circuits to manage the flow of air, regulate temperature, and ensure comfortable and safe indoor environments.

20.1 TYPICAL CIRCUITS FOUND IN HEATING AND VENTILATION SYSTEMS

20.1.1 Heating System Circuits

Heating systems are designed to generate and distribute heat throughout a space. They include several circuits that work together to maintain a consistent and comfortable temperature.

a) Heating Control Circuit

- **Description** – The control circuit manages the overall operation of the heating system. It includes the thermostat, relays, and control board.
- **Components**
  - **Thermostat** – A device that monitors room temperature and sends signals to the control circuit to turn the heating system on or off based on the set temperature.
  - **Control Board** – The brain of the heating system that processes signals from the thermostat and other sensors, sending commands to various components.
  - **Relays** – Electrically operated switches that activate components such as the burner, blower, or heating elements.
- **Operation** – When the thermostat detects a temperature below the set point, it sends a signal to the control board, which closes the relay, activating the heating components (e.g., gas valve, electric heating element, or blower motor) to produce heat.

b) Burner or Heat Generation Circuit

- **Description** – This circuit is responsible for generating heat in systems like gas or oil furnaces.
- **Components**
  - **Burner** – A device that mixes fuel (gas or oil) with air and ignites it to generate heat.

  - **Igniter** – A component that lights the fuel mixture, typically using a spark or a hot surface igniter.
  - **Gas Valve/Oil Pump** – Controls the flow of fuel to the burner.
  - **Flame Sensor** – Ensures the burner is properly ignited and shuts off the fuel supply if no flame is detected.
- **Operation** – The control circuit activates the gas valve or oil pump, allowing fuel to flow to the burner. The igniter lights the fuel, generating heat, which is transferred to the air or water in the system.

c) Blower/Fan Circuit

- **Description** – The blower or fan circuit distributes heated air throughout the building.
- **Components**
  - **Blower Motor** – A motor-driven fan that circulates air over the heat exchanger or heating elements.
  - **Limit Switch** – A safety device that turns off the blower if the system overheats.
- **Operation** – Once the heat exchanger reaches the desired temperature, the blower motor is activated, forcing air through the ducts and into the living space.

d) Safety Circuit

- **Description** – This circuit ensures the heating system operates safely, preventing accidents or damage.
- **Components**
  - **High Limit Switch** – Shuts down the system if the temperature exceeds a safe limit.
  - **Pressure Switch** – Monitors proper airflow or venting and prevents the burner from operating if air pressure is not adequate.
  - **Carbon Monoxide Detector** – In systems using combustion, it detects dangerous levels of carbon monoxide and shuts down the system if necessary.

20.1.2 Ventilation System Circuits

Ventilation systems are designed to move fresh air into a building and expel stale air, ensuring a healthy indoor environment. Key circuits in ventilation systems include:

a) Air Handling Circuit

- **Description** – This circuit manages the intake, conditioning, and distribution of air in a building.
- **Components**
  - **Air Handler** – A central unit that contains a fan, filter, and possibly heating or cooling elements.
  - **Fan Motor** – Powers the fan to circulate air throughout the ducts.
  - **Dampers** – Adjustable plates that regulate airflow within the duct system.
  - **Air Filters** – Remove dust and contaminants from the incoming air.
- **Operation** – The air-handling unit pulls fresh air from the outside or recirculates indoor air, filters it, and conditions it (heating or cooling) before distributing it through the duct system.

b) Exhaust Circuit

- **Description** – This circuit is responsible for removing stale, contaminated, or humid air from the building.
- **Components**
  - **Exhaust Fan** – A fan that pulls air out of the building, often located in bathrooms, kitchens, or industrial areas.
  - **Ductwork** – Channels that guide the stale air from the interior to the exterior of the building.
  - **Vent Hood** – A cover that protects the exhaust outlet from weather and prevents backflow.
- **Operation** – The exhaust fan creates a negative pressure, drawing air from inside the building and expelling it outside. This may operate continuously or intermittently based on occupancy or humidity levels.

c) Fresh Air Intake Circuit

- **Description** – This circuit brings fresh outdoor air into the ventilation system.
- **Components**
  - **Intake Fan** – A fan that draws outdoor air into the system.
  - **Filters** – Clean the incoming air to remove pollutants and particles.

- **Energy Recovery Ventilator (ERV)** or **Heat Recovery Ventilator (HRV)** – Exchanges heat or moisture between the incoming fresh air and outgoing stale air to improve efficiency.

- **Operation** – The intake fan pulls in outdoor air, which passes through filters. An ERV or HRV can pre-condition the air before it's circulated indoors.

d) Temperature Control Circuit (in HVAC Systems)

- **Description** – In systems combining heating and ventilation (HVAC), this circuit controls temperature adjustments for heating and cooling.
- **Components**
  - **Thermostat** – Controls temperature settings for both heating and cooling.
  - **Sensors** – Detect air temperature, humidity, and pressure.
  - **Control Board** – Manages signals from the thermostat to adjust heating, cooling, and fan settings.
- **Operation** – The thermostat sends signals to either the heating or cooling systems to achieve the desired temperature. This circuit adjusts the fan speed and damper positions to balance temperature and airflow.

20.1.3 Examples of Application in Different Systems

1 Residential HVAC System

- Uses a combination of heating circuits (for furnaces) and ventilation circuits (air handlers) to regulate indoor climate, often using a central thermostat to control both heating and ventilation needs.

2 Industrial Ventilation System

- Incorporates specialized exhaust circuits to remove contaminants or fumes in manufacturing areas, with dedicated intake circuits to maintain air quality and balance.

3 Commercial Heating System

- Employs complex blower and air handling circuits to ensure uniform heating across large spaces like shopping centers or office buildings, often using dampers to control different zones.

20.1.4 Summary

In summary, heating and ventilation systems rely on several integrated circuits to function efficiently. Understanding how each circuit operates allows for better design, maintenance, and troubleshooting, ensuring that buildings remain comfortable and safe.

20.2 BASIC OPERATION OF HEATING AND VENTILATION SYSTEMS

Heating and ventilation systems are essential for maintaining comfortable and safe indoor environments in homes, offices, and industrial spaces. These systems work to regulate temperature, improve air quality, and ensure proper air circulation.

20.2.1 Heating Systems

Heating systems are designed to increase the temperature of indoor spaces to ensure comfort during cold weather. The primary components include a heat source, distribution system, and control mechanisms.

1 **Heat Generation**

- **Furnace (Gas or Oil)** – Burns fuel to generate heat.
- **Boiler** – Heats water or steam, which is circulated through radiators or a radiant floor system.
- **Electric Heater** – Uses electrical resistance to generate heat.
- **Heat Pump** – Extracts heat from the outside air or ground and transfers it indoors.

2 **Heat Distribution**

- **Air Ducts** – If using a furnace, heated air is pushed through a network of ducts by a blower fan. The warm air is delivered to each room through vents.
- **Radiators** – In a boiler system, hot water or steam travels through pipes to radiators, which then release heat into the room.
- **Baseboard Heaters** – These are electric or hot water heating units mounted along baseboards to radiate heat directly.

3 **Temperature Control**

- **Thermostat** – A thermostat monitors the indoor temperature and turns the heating system on or off based on the set temperature. It ensures that the room temperature is maintained at the desired comfort level.

- **Sensors** – Some advanced systems include additional sensors that monitor outdoor temperature or humidity levels for more precise control.

20.2.2 Ventilation Systems

Ventilation systems are designed to control indoor air quality by introducing fresh air, removing stale or contaminated air, and maintaining proper humidity levels. The primary components include fans, ducts, filters, and control mechanisms. Here's how they operate:

1 **Fresh Air Intake**

- **Air Handling Unit (AHU)** – Pulls in fresh outdoor air through intake vents. The AHU may include filters to remove dust, pollen, and other contaminants from the incoming air.
- **Energy Recovery Ventilator (ERV)** or **Heat Recovery Ventilator (HRV)** – Transfers heat or humidity between incoming and outgoing air streams to improve efficiency and maintain indoor comfort.

2 **Air Circulation**

- **Ducts** – Once the fresh air is filtered, it is circulated through a system of ducts to different rooms.
- **Fans/Blowers** – Circulate air through the ducts and maintain airflow throughout the building. Fans can be adjusted for speed and volume to ensure proper circulation.

3 **Exhaust**

- **Exhaust Fans** – Located in bathrooms, kitchens, or industrial areas to remove stale, moist, or contaminated air. This prevents buildup of humidity, odors, and pollutants.
- **Exhaust Ducts** – Carry the expelled air out of the building, ensuring that fresh air can replace it.

4 **Humidity Control**

- **Dehumidifier** – Reduces indoor humidity by removing excess moisture from the air.
- **Humidifier** – Adds moisture to the air if it is too dry, especially during winter when heating can lower indoor humidity.
- Proper humidity control prevents problems like mold growth, static electricity, or discomfort from dry air.

20.2.3 Integrated Heating and Ventilation (HVAC) Systems

In many buildings, heating and ventilation systems are combined into an **HVAC** (Heating, Ventilation, and Air Conditioning) system that controls temperature, air quality, and comfort all at once.

20.2.4 Basic Operation of HVAC Systems

- **Heating Mode** – When heating is needed, the system draws in cold indoor air, heats it using the furnace, boiler, or heat pump, and then distributes the warm air through the ducts.
- **Ventilation Mode** – The system brings in fresh outdoor air, filters it, and distributes it indoors. It simultaneously expels stale indoor air to maintain air quality.
- **Air Conditioning Mode** – In summer, the HVAC system reverses its operation. It cools the indoor air using refrigerants and removes excess moisture, ensuring comfortable humidity levels.
- **Thermostat Control** – An integrated thermostat monitors the indoor temperature and controls the heating, ventilation, and air conditioning components to maintain the desired conditions.

20.2.5 Summary

In summary, heating and ventilation systems are crucial for creating comfortable and safe indoor environments by regulating temperature, humidity, and air quality. They operate through a combination of heat generation, air distribution, fresh air intake, and exhaust systems, all controlled by thermostats and sensors. Understanding the basics of these systems helps with maintenance, troubleshooting, and optimizing efficiency.

20.3 BASIC COMPONENTS OF HEATING AND VENTILATION SYSTEMS

Heating and ventilation systems have a variety of components that work together to control indoor temperature, air quality, and overall comfort. These components can be broadly categorized based on their function—such as heating, air distribution, and ventilation control. The **basic components** and **methods of control** found in typical heating and ventilation systems is as stated below:

20.3.1 Components in Sections of Heating and Ventilation Systems

1 Heat Generation Components

- **Furnace** – Burns fuel (gas, oil, or propane) to generate heat, which is then distributed throughout the building.

- **Boiler** – Heats water or produces steam, which circulates through radiators or underfloor heating systems to provide warmth.
- **Heat Pump** – Transfers heat from outside (air, ground, or water) to inside, functioning as both a heating and cooling unit.
- **Electric Heater** – Uses electrical resistance elements to generate heat directly, often found in baseboard heaters or portable space heaters.

2 **Air Distribution Components**

- **Ducts** – A network of tubes (usually made of sheet metal or flexible material) that carry heated or cooled air from the heating source to different rooms.
- **Registers/Grilles** – Vents that allow air to enter or exit rooms, usually adjustable to control airflow.
- **Blower Fan** – Circulates air through ducts, pushing warm or cool air throughout the building.
- **Radiators** – Transfer heat from hot water or steam into the room, usually connected to a boiler system.
- **Diffusers** – Distribute air evenly in a room, often used in commercial or industrial HVAC systems to maintain consistent airflow.

3 **Ventilation Components**

- **Air Handling Unit (AHU)** – Contains fans, filters, heating/cooling coils, and dampers for circulating and conditioning air.
- **Exhaust Fans** – Remove stale, moist, or polluted air from specific areas like kitchens or bathrooms.
- **Ventilation Ducts** – Carry fresh air from outside into the building and exhaust stale air to the outside.
- **Filters** – Remove dust, pollen, and other particles from the air to improve indoor air quality. These can be disposable, washable, or high-efficiency filters like HEPA.
- **Dampers** – Control the flow of air in the duct system. They can be adjusted manually or automatically to balance airflow.

4 **Control Components**

- **Thermostat** – The primary control device for temperature, allowing users to set the desired indoor temperature. It turns heating or cooling systems on or off based on temperature readings.
- **Humidistat** – A device that monitors and controls humidity levels within a space, often integrated with humidifiers or dehumidifiers.
- **Timers/Programmable Controllers** – Allow heating and ventilation systems to be scheduled, saving energy by operating only when needed (e.g., programmable thermostats).
- **Sensors** – Measure temperature, humidity, or air quality. These are often used in advanced systems to provide real-time adjustments to maintain comfort.
- **Zoning Controls** – Use dampers, thermostats, and sensors to create different temperature zones within a building, allowing for individual control of specific areas.
- **Variable Speed Drives (VSDs)** – Control the speed of fans or pumps to regulate airflow and save energy. They are used in advanced HVAC systems for more precise control.

20.3.2 Methods of Control in Heating and Ventilation Systems

1 **Manual Control**

- Users manually adjust settings on components like thermostats, dampers, and fans to control the temperature and airflow.
- Simple manual switches are often used for operating exhaust fans, controlling individual room heaters, or opening/closing vents.

2 **Thermostatic Control**

- **Thermostat-Based Systems** – These systems use a thermostat to monitor the indoor temperature. When the temperature falls below the set point, the heating system is activated. If it rises above the set point, the system turns off.
- **Programmable Thermostats** – Allow users to set different temperatures for different times of the day or days of the week, optimizing energy usage.

3 **Zoning Control**

- **Zoned Systems** – Use multiple thermostats and dampers to control the temperature in different parts of a building independently. This reduces energy costs by heating or cooling only occupied areas.

- Each zone can be managed separately, offering customized comfort levels for various areas of a building.

4 **Automatic Control**

- **Sensors** – Systems equipped with sensors can automatically adjust based on real-time data. For example, a carbon dioxide sensor may trigger increased ventilation if indoor CO_2 levels get too high.
- **Smart Thermostats** – Learn user behavior and adjust settings automatically. They often include Wi-Fi connectivity, allowing remote adjustments through a smartphone app.

5 **Remote and Smart Control**

- **Remote Controls** – Handheld devices or wall-mounted panels that allow users to control the heating and ventilation system without accessing the main unit.
- **Smart Systems** – Use internet-connected devices, allowing for remote operation via smartphone apps or integration with home automation systems like Google Home or Amazon Alexa.
- **Learning Algorithms** – Some advanced smart systems learn patterns of use and optimize heating/cooling schedules automatically to reduce energy consumption.

6 **Proportional Control**

- In sophisticated HVAC systems, proportional control allows for precise adjustments. For example, the system will not just turn on or off but adjust the intensity based on the difference between the current temperature and the desired set point.

20.3.3 Examples of Control Application

- **Example 1** – In a basic residential setup, a simple thermostat turns the furnace on when the room temperature drops below 20°C and turns it off when it reaches 22°C.
- **Example 2** – In a commercial building, a smart HVAC system equipped with zoning and VSD controls will adjust airflow based on occupancy sensors. If no one is in a particular zone, dampers close to save energy, and fans reduce speed to decrease power consumption.
- **Example 3** – In industrial environments, AHUs equipped with sensors will increase ventilation if indoor pollutants exceed safe levels, automatically adjusting airflow to maintain air quality.

20.3.4 Conclusion

By understanding the basic components and methods of control, users can better manage heating and ventilation systems for efficiency, comfort, and safety.

CHAPTER TWENTY ONE

21.0 AIR CONDITIONING AND REFRIGERATION SYSTEM CIRCUITS

Air conditioning and refrigeration systems rely on specific circuits to control the cooling and conditioning of air. These circuits include various components that manage the flow of refrigerant, regulate temperature, and ensure efficient operation.

21.1 KEY CIRCUITS IN AIR CONDITIONING AND REFRIGERATION SYSTEMS

21.1.1 Air Conditioning and Refrigeration System Circuits

1 Refrigeration Circuit

The refrigeration circuit is the core of any air conditioning or refrigeration system. It involves a closed loop where a refrigerant is circulated to absorb and remove heat. Here are the main components of the refrigeration circuit:

- **Compressor** – Acts as the "heart" of the system, compressing the refrigerant gas and increasing its pressure. This raises the refrigerant's temperature, making it a hot, high-pressure gas.
- **Condenser** – A coil or heat exchanger where the high-pressure refrigerant releases its heat to the surroundings (usually air or water). As it loses heat, it condenses into a high-pressure liquid.
- **Expansion Valve** – A valve or orifice that reduces the pressure of the refrigerant as it moves through, causing it to expand and cool rapidly. This results in a low-pressure, low-temperature mixture of liquid and gas.
- **Evaporator** – A coil or heat exchanger where the cold refrigerant absorbs heat from the air (in air conditioners) or from stored goods (in refrigeration units). This causes the refrigerant to evaporate into a gas, completing the circuit.

Example – In a household refrigerator, the compressor pumps refrigerant through the system, where it absorbs heat from the food compartment in the evaporator and releases it outside through the condenser coils at the back of the fridge.

2 Electrical Control Circuit

The **electrical control circuit** manages the operation of the refrigeration components and ensures that the system operates efficiently and safely. Key components include:

- **Thermostat** – A temperature-sensing device that controls the compressor by monitoring the temperature. When the temperature rises above a set point, it activates the compressor to start cooling. When the desired temperature is reached, it turns off the compressor.
- **Control Board** – The "brain" of the system, which coordinates various signals from sensors, thermostats, and timers. It manages when the compressor, fans, and other components should operate.
- **Relays and Contactors** – Switches that control the flow of electricity to the compressor, fans, and other motorized parts. They are activated by signals from the control board or thermostat.
- **Defrost Timer (in refrigeration)** – A component that periodically initiates a defrost cycle to remove frost buildup on the evaporator coils, ensuring efficient heat transfer.

Example: In a central air conditioning system, the thermostat signals the control board to activate the compressor and fan when the room temperature rises above the set point. If a defrost cycle is needed in a refrigeration unit, the defrost timer triggers heaters to melt any ice buildup on the evaporator.

3 Air Flow Circuit (in Air Conditioning Systems)

This circuit handles the movement of air across the cooling coil and into the space that needs to be conditioned. It includes:

- **Blower Fan** – Moves air over the evaporator coils, allowing the air to be cooled before being distributed throughout the room or building.
- **Ductwork** – A network of pathways that direct the cooled air from the air handler to different parts of a building. Return ducts bring warm air back to the system for re-cooling.
- **Air Filters** – Remove dust, dirt, and other particles from the air before it is circulated, improving indoor air quality and protecting the evaporator coils from debris.
- **Dampers** – Devices installed in ductwork that regulate airflow, allowing for temperature zoning within a building.

Example – In a residential HVAC system, the blower fan circulates warm indoor air through the filter, then over the evaporator coil where it is cooled. The cool air is then delivered back to living spaces through the ducts.

4 **Refrigerant Flow Circuit (in Split AC Systems)** Split air conditioning systems have both indoor and outdoor units, connected by refrigerant lines. The refrigerant flow circuit includes:

- **Indoor Unit (Evaporator)** – Contains the evaporator coil and fan, absorbing heat from the room air.
- **Outdoor Unit (Condenser)** – Houses the compressor and condenser coil, releasing absorbed heat to the outdoor environment.
- **Copper Tubing** – Connects the indoor and outdoor units, facilitating the transfer of refrigerant. This tubing consists of a larger **suction line** (carrying cool gas back to the compressor) and a smaller **liquid line** (carrying high-pressure liquid refrigerant to the evaporator).

Example – In a split AC unit, the indoor evaporator absorbs heat from the room. The warmed refrigerant is sent to the outdoor unit via copper tubing, where the condenser dissipates the heat.

21.1.2 Operational Steps of the Cooling Cycle

The basic cooling cycle for air conditioning and refrigeration systems can be broken down as follows:

1 **Compression** – The refrigerant gas is compressed by the compressor, increasing its temperature and pressure.

2 **Condensation** – The hot gas passes through the condenser, where it releases heat and condenses into a liquid.

3 **Expansion** – The liquid refrigerant flows through the expansion valve, dropping in pressure and temperature.

4 **Evaporation** – The cold, low-pressure refrigerant absorbs heat in the evaporator, cooling the air (in AC) or preserving stored items (in refrigeration).

21.1.3 Types of Circuits in Air Conditioning and Refrigeration Systems

- **Single-Stage Circuit** – Uses a single compressor that cycles on and off based on temperature demands. It is commonly found in simple residential or small commercial systems.
- **Multi-Stage Circuit** – Features multiple compressors or variable speed compressors for better efficiency, providing cooling at varying levels based on need. This is often used in larger commercial systems.

- **Heat Pump Circuit** – A special type of circuit where the refrigeration cycle can be reversed to provide both heating and cooling. The reversing valve switches the direction of refrigerant flow depending on the mode (heating or cooling).

21.1.4 Summary

By understanding these circuits, maintenance and troubleshooting of air conditioning and refrigeration systems become more manageable, as each component plays a crucial role in the system's overall efficiency and functionality.

21.2 KEY COMPONENTS IN AIR CONDITIONING AND REFRIGERATION SYSTEMS

The basic operation of air conditioning and refrigeration systems relies on the principles of heat transfer and the phase changes of a refrigerant. These systems are designed to remove heat from an enclosed space, creating a cooler environment.

21.2.1 Breakdown of how Air Conditioning and Refrigeration Systems Operate

- **Compressor** – Pumps the refrigerant through the system. It compresses the refrigerant gas, increasing its pressure and temperature.
- **Condenser Coil** – Releases the absorbed heat from the refrigerant to the external environment. The refrigerant changes from a high-pressure gas to a high-pressure liquid as it cools down.
- **Expansion Valve/Capillary Tube** – Controls the flow of refrigerant into the evaporator. It reduces the pressure of the liquid refrigerant, causing it to cool down rapidly.
- **Evaporator Coil** – Absorbs heat from the air or contents being cooled. The refrigerant evaporates, turning from a low-pressure liquid to a low-pressure gas as it absorbs heat.
- **Refrigerant** – The working fluid that cycles through the system, absorbing and releasing heat. It undergoes phase changes (liquid to gas and back) to facilitate cooling.

21.2.2 Four Key Stages of Air Conditioning and Refrigeration Cycle

a) Compression

- The process begins with the **compressor**, which takes in low-pressure, low-temperature refrigerant gas from the evaporator.
- The compressor compresses the refrigerant, raising its pressure and temperature. This transforms the refrigerant into a hot, high-pressure gas.

- The hot gas is then pushed through the system to the next stage.

b) Condensation

- The high-pressure, hot refrigerant gas flows into the **condenser coil**, usually located outside the space to be cooled.
- In the condenser, the refrigerant releases the absorbed heat to the surroundings (usually air or water). As it loses heat, the refrigerant condenses into a high-pressure liquid.
- A fan may be used to help dissipate the heat from the condenser coil more effectively.

c) Expansion

- The high-pressure liquid refrigerant passes through the **expansion valve** or capillary tube. This device reduces the refrigerant's pressure, causing it to cool down significantly.
- As the pressure drops, the refrigerant becomes a cold, low-pressure liquid, preparing it for the next phase where it can absorb heat.

d) Evaporation

- The cooled, low-pressure liquid refrigerant enters the **evaporator coil**, which is located in the space that needs to be cooled (e.g., inside a room or refrigerator compartment).
- In the evaporator, the refrigerant absorbs heat from the surrounding air or content, causing it to evaporate and turn back into a gas.
- As it absorbs heat, the air or content in contact with the evaporator cools down.
- A fan blows air across the evaporator coil to distribute the cooled air throughout the space.
- The cycle then repeats, with the refrigerant gas returning to the compressor to start the process over again.

21.2.3 Air Conditioning vs. Refrigeration Operation

- **Air Conditioning Systems** – These systems are designed to cool and condition indoor air, often providing dehumidification and filtering as well. The cooled air is circulated through ductwork or directly into a room using fans, keeping indoor temperatures comfortable.
- **Refrigeration Systems** – These systems are used for cooling and preserving food or other perishable items. They are commonly found in refrigerators, freezers, and cold storage facilities. Unlike air conditioners, refrigeration systems maintain lower temperatures for prolonged periods, focusing more on preserving the quality of stored items.

21.2.4 Cooling Cycle in a Real-World Example

Example: A Split Air Conditioner

- The indoor unit (evaporator) absorbs heat from the room air, cooling the space.
- The refrigerant carries the absorbed heat to the outdoor unit (condenser), where it is expelled outside.
- The compressor, typically located in the outdoor unit, pumps the refrigerant through the cycle.
- A thermostat monitors room temperature and signals the system to cycle on or off to maintain the desired comfort level.

Example: A Household Refrigerator

- Inside the refrigerator, the evaporator absorbs heat from the air, keeping food items cool.
- The compressor, located at the back or bottom of the unit, pumps the refrigerant.
- The condenser releases the heat absorbed inside the refrigerator to the outside air, typically behind or underneath the appliance.

21.2.5 Additional Functions in Air Conditioning Systems

- **Dehumidification** – Air conditioning systems often include a dehumidification function, where moisture from the air condenses on the evaporator coils, reducing indoor humidity.
- **Air Filtration** – Many air conditioning units have filters to capture dust, allergens, and other particles, improving indoor air quality.
- **Heating Capability (Heat Pumps)** – Some air conditioners have a reverse cycle (heat pump), allowing them to provide heating by reversing the refrigerant flow, absorbing heat from the outside air, and releasing it indoors.

21.2.6 Control Systems

- **Thermostat** – Regulates the temperature by controlling the compressor and fans based on the desired set temperature.
- **Sensors** – Monitor the temperature of the air or refrigerant, providing feedback to the control unit for adjustments.
- **Timers and Programmable Controls** – Allow users to set specific cooling or heating schedules, optimizing energy use.

21.2.7 Summary

In summary, the basic operation of air conditioning and refrigeration systems revolves around the refrigeration cycle, which continuously absorbs and removes heat from a space to maintain the desired temperature. Proper management of refrigerant flow and heat transfer, along with efficient control systems, ensures that these systems function effectively, providing comfort and preserving.

21.3 BASIC OPERATION OF AIR CONDITIONING AND REFRIGERATION SYSTEMS

The basic operation of air conditioning and refrigeration systems primarily relies on refrigerants as the working fluid, along with other supportive fluids for system operation.

21.3.1 Refrigerants

- **Role** – Refrigerants are the primary fluids that absorb and release heat as they cycle through the system, facilitating cooling. They change states (from liquid to gas and back) during the process to absorb and expel heat.
- **Common Types of Refrigerants**
  - ***R-410A*** – A common refrigerant in modern air conditioning systems, known for being environmentally friendly compared to older options.
  - ***R-22 (Freon)*** – An older refrigerant, less common today due to environmental concerns (phased out in many countries).
  - ***R-134a*** – Widely used in refrigeration systems, particularly in automotive air conditioning and domestic refrigerators.
  - ***R-404A*** – Often used in commercial refrigeration systems, like supermarket freezers.
  - ***R-32*** – Known for higher efficiency and lower global warming potential, increasingly used in modern air conditioners.

21.3.2 Lubricating Oils

- **Role** – Used to lubricate the moving parts of the compressor in air conditioning and refrigeration systems, reducing friction and wear. The lubricating oil mixes with the refrigerant and circulates throughout the system.
- **Common Types**
  - **Mineral Oil** – Often used with older refrigerants like R-22.

- **Polyol Ester (POE) Oil** – Commonly used with newer refrigerants like R-410A and R-134a.
- **Alkylbenzene (AB) Oil** – Sometimes used in systems with specific refrigerant types.

21.3.3 Water (in Certain Systems)

- **Role**: In some air conditioning systems, particularly **water-cooled systems**, water is used as a secondary cooling fluid. It helps transfer heat from the refrigerant to the external environment.
- **Examples**
  - ***Chilled Water Systems*** – Use water as a cooling medium, circulating chilled water to cool spaces.
  - ***Cooling Towers*** – Use water to dissipate heat from refrigerant loops in large-scale HVAC systems.

21.3.4 Antifreeze/Coolants (in Certain Systems)

- **Role** – In some air conditioning or refrigeration systems, antifreeze solutions are used to prevent freezing in extreme conditions. These fluids are mixed with water in systems where sub-zero temperatures are possible.
- **Common Types**
  - ***Ethylene Glycol*** – A commonly used antifreeze in refrigeration systems.
  - ***Propylene Glycol*** – An alternative antifreeze, often used in food-related applications due to its lower toxicity.

21.3.5 Summary

- **Refrigerants** are the core fluids in both air conditioning and refrigeration systems, responsible for the heat transfer cycle.
- **Lubricating oils** are crucial for maintaining the efficiency and longevity of compressors.
- In **water-cooled** and **chilled water systems**, water acts as an additional heat-transfer fluid.
- **Antifreeze** fluids are used in specific systems to manage extreme temperatures.

These fluids work together to create an efficient and effective cooling cycle in air conditioning and refrigeration systems, ensuring comfort and preservation.

CHAPTER TWENTY TWO

22.0 DESIGN OF NORMAL (MANUAL) AND AUTOMATIC AIR CONDITIONING SYSTEMS

The design of **normal (manual)** and **automatic air conditioning systems** varies significantly in terms of control, functionality, and user interaction.

22.1 DIFFERENCES BETWEEN NORMAL AND AUTOMATIC AIR CONDITIONING

22.1.1 Temperature Control

- **Normal Air Conditioning System**
  - **Manual Adjustment** – Users set the temperature manually using a dial, switch, or basic digital display.
  - **On/Off Operation** – Once the desired temperature is reached, the unit may simply turn off the compressor, requiring manual adjustment to change settings.
  - **Limited Precision** – Temperature control is less precise, often leading to fluctuations as the system cycles on and off.
- **Automatic Air Conditioning System**
  - **Self-Regulating** – Uses a thermostat with sensors to automatically adjust and maintain a set temperature without user intervention.
  - **Continuous Adjustment** – Automatically modulates the system's performance to maintain a consistent temperature.
  - **Greater Accuracy** – Provides more precise temperature control, minimizing fluctuations.

22.1.2 Fan and Airflow Control

- **Normal Air Conditioning System**
  - **Fixed Fan Speeds** – Fan speed is manually controlled, with options like low, medium, or high.
  - **No Automatic Adjustment** – The user must manually change fan speed to adjust airflow based on comfort.
  - **Constant Airflow** – Air distribution does not change unless manually altered by the user.

- **Automatic Air Conditioning System**
  - **Variable Fan Speeds** – The system automatically adjusts fan speed based on the temperature difference and room requirements.
  - **Adaptive Airflow** – Adjusts air distribution to maintain comfort, often including features like automatic vent adjustment.
  - **Energy Efficiency** – The system optimizes fan speed for energy efficiency and noise reduction.

22.1.3 Modes of Operation

- **Normal Air Conditioning System**
  - **Limited Modes** – Typically includes basic modes such as cooling, heating, or fan-only, selected manually by the user.
  - **User-Dependent** – Requires manual intervention to change modes or adjust settings.
- **Automatic Air Conditioning System**
  - **Advanced Modes** – Includes modes like "Auto," "Eco," "Sleep," or "Dehumidify," which adjust settings automatically based on environmental conditions.
  - **Smart Sensing** – Uses sensors to detect indoor conditions and switches between modes without user input.

22.1.4 Control Interface

- **Normal Air Conditioning System**
  - **Basic Controls –** Typically features simple controls like knobs, buttons, or basic digital displays.
  - **No Remote Access** – Adjustments must be made directly at the unit; limited or no remote functionality.
- **Automatic Air Conditioning System**
  - **Advanced Controls** – Features digital interfaces with programmable thermostats, touchscreens, or smart displays.
  - **Remote Access –** Can be controlled remotely through mobile apps, smart devices, or wireless remotes.

22.1.5 Energy Efficiency

- **Normal Air Conditioning System**
  - **Higher Energy Usage –** Due to manual operation, the unit may run longer than necessary, leading to increased energy consumption.
  - **On/Off Operation** – The compressor cycles on and off, which can be less energy-efficient.
- **Automatic Air Conditioning System**
  - **Energy-Saving Features** – Often includes variable-speed compressors and fan motors that adjust automatically to save energy.
  - **Optimized Performance –** Maintains energy efficiency by adjusting the system based on real-time conditions, reducing overall consumption.

22.1.6 Sensors and Automation

- **Normal Air Conditioning System**
  - **Basic Thermostat** – Utilizes a simple thermostat for temperature sensing, requiring manual adjustments.
  - **No Advanced Sensors –** Typically lacks additional sensors, limiting its ability to adapt to changes in the environment.
- **Automatic Air Conditioning System**
  - **Advanced Sensors –** Uses multiple sensors to monitor temperature, humidity, and occupancy, allowing the system to adapt automatically.
  - **Smart Technology** – Includes features like occupancy sensors, light sensors, and humidity sensors to optimize performance.

22.1.7 User Experience

- **Normal Air Conditioning System**
  - **User-Dependent –** Requires constant user input to manage settings and comfort.
  - **Manual Control** – Lacks the ability to automate comfort adjustments, placing more burden on the user.
- **Automatic Air Conditioning System**
  - **User Convenience** – Offers a "set and forget" experience, where the system maintains optimal conditions without ongoing user interaction.

- **User Profiles** – Some systems allow programming of personalized settings or schedules based on user preference.

22.1.8 Cost and Complexity

- **Normal Air Conditioning System**
  - **Lower Initial Cost** – Generally cheaper to purchase and install due to simpler components and controls.
  - **Less Complex** – Easier to understand and operate with minimal features.
- **Automatic Air Conditioning System**
  - **Higher Initial Cost** – More expensive due to advanced technology, sensors, and features.
  - **Increased Complexity** – More complex to set up and may require professional installation and maintenance.

22.1.8.1 Table of Summary Table for Normal AC and Automatic AC

Feature	Normal AC	Automatic AC
Temperature Control	Manual adjustment	Self-regulating, precise
Fan Speed	Fixed, manual control	Variable, automatic adjustment
Modes of Operation	Basic modes (cooling, heating, etc.)	Advanced modes (Eco, Auto, Sleep)
Control Interface	Basic knobs and buttons	Digital, smart controls, remote access
Energy Efficiency	Higher energy use, less efficient	Optimized for energy savings
Sensors	Simple thermostat	Multiple sensors (temperature, humidity)
User Experience	Manual, user-dependent	Automated, user-friendly
Cost	Lower	Higher
Complexity	Simpler design	Advanced, complex

These differences reflect the balance between simplicity and control in normal systems versus the automation and efficiency provided by automatic systems.

22.2 FAULTS IN HEATING, VENTILATION, AND AIR CONDITIONING (HVAC) SYSTEMS

Faults in heating, ventilation, and air conditioning (HVAC) systems can significantly affect performance, energy efficiency, and comfort.

22.2.1 Common Faults in Heating Systems

1 **No Heat or Insufficient Heat**

- ***Symptoms*** – The system does not turn on, or it runs without adequately heating the space.
- **Possible Causes**
  - Faulty thermostat (incorrect settings or malfunction).
  - Tripped circuit breaker or blown fuse.
  - Pilot light or ignition control problems.
  - Clogged or dirty air filters.
  - Malfunctioning heat exchanger or burner.

2 **Uneven Heating**

- ***Symptoms*** – Some rooms are warmer than others, or there are noticeable cold spots.
- **Possible Causes**
  - Poorly balanced ductwork or closed vents.
  - Blocked or dirty air filters.
  - Incorrect thermostat location (affected by sunlight or drafts).
  - Issues with dampers not opening or closing properly.

3 **Unusual Noises**

- ***Symptoms*** – Rattling, banging, whistling, or squealing sounds during operation.
- **Possible Causes**
  - Loose components such as belts, bearings, or panels.
  - Blower motor or fan problems.
  - Air in the heating system (common in water-based systems).
  - Worn-out or malfunctioning furnace parts (e.g., blower fan or burners).

4 **Frequent Cycling (Short-Cycling)**

- *Symptoms* – The heating system turns on and off too frequently.
- **Possible Causes**
  - Oversized furnace or heating unit.
  - Dirty or clogged air filters restricting airflow.
  - Faulty thermostat sensor or incorrect thermostat placement.
  - Blocked or closed vents.

5 **High Energy Bills**

- *Symptoms* – Unexpected spikes in energy consumption.
- **Possible Causes**
  - Leaky or poorly insulated ductwork.
  - Dirty or blocked filters causing the system to work harder.
  - Inefficient or outdated heating equipment.
  - Malfunctioning components (e.g., thermostat, blower fan).

22.2.2 Common Faults in Ventilation Systems

1 **Poor Airflow**

- *Symptoms* – Reduced air movement from vents, uneven distribution.
- **Possible Causes**
  - Blocked or dirty filters.
  - Obstructed air ducts or vents.
  - Fan motor malfunction or wear.
  - Leaks in the ductwork reducing pressure.

2 **Strange Odors**

- *Symptoms:* Musty, burning, or foul smells coming from vents.
- **Possible Causes**
  - Mold or mildew in the ductwork or filters.
  - Accumulation of dust and debris in the system.

- Burnt-out or overheated components.
- Animal infestation (pests or rodents) in the ductwork.

3 Excessive Dust or Allergens

- ***Symptoms***: Dust accumulation near vents or increased allergy symptoms.
- **Possible Causes**
  - Dirty filters not capturing contaminants.
  - Air leaks in ducts bringing in unfiltered air.
  - Inadequate filtration or poorly maintained air purification systems.

4 Inadequate Ventilation

- ***Symptoms***: Rooms feel stuffy, humid, or have poor air quality.
- **Possible Causes**
  - Insufficient fan speed or inadequate fan capacity.
  - Blocked or obstructed intake or exhaust vents.
  - Faulty or improperly sized exhaust fans.
  - Malfunctioning dampers that regulate airflow.

22.2.3 Common Faults in Air Conditioning Systems

1 AC Not Cooling or Insufficient Cooling

- ***Symptoms***: The system is running, but the air is not cool, or it takes a long time to cool the room.
- **Possible Causes**
  - Low refrigerant levels or refrigerant leaks.
  - Dirty condenser or evaporator coils reducing efficiency.
  - Clogged air filters restricting airflow.
  - Thermostat issues (incorrect settings or malfunction).
  - Compressor failure or electrical problems.

2 AC Unit Leaking Water

- ***Symptoms***: Water pooling around the indoor unit or dripping.

- **Possible Causes**
  - Clogged condensate drain line.
  - Frozen evaporator coils causing excessive condensation.
  - Low refrigerant levels leading to coil freezing.
  - Broken or disconnected drain pan.

3 **Unusual Noises**

- ***Symptoms:*** Grinding, rattling, buzzing, or squealing noises from the indoor or outdoor unit.
- **Possible Causes**
  - Loose parts or screws.
  - Fan motor problems (worn bearings or damaged blades).
  - Compressor issues, including failure or wear.
  - Refrigerant line leaks causing hissing sounds.

4 **AC Constantly Running or Short-Cycling**

- ***Symptoms:*** The air conditioner never turns off or turns on and off rapidly.
- **Possible Causes**
  - Oversized or undersized AC unit for the space.
  - Blocked airflow due to dirty filters or blocked vents.
  - Malfunctioning thermostat or faulty wiring.
  - Refrigerant problems causing the system to work harder.

5 **Warm Air Blowing from Vents**

- ***Symptoms***: Air conditioning is on, but warm or room temperature air is coming out.
- **Possible Causes**
  - Low refrigerant levels or leak.
  - Dirty condenser or evaporator coils.
  - Reversing valve malfunction in heat pump systems.
  - Compressor issues or fan motor failure.

22.2.4 Diagnostic and Repair Tips

1 Visual Inspection

- Check for visible signs of damage (e.g., loose parts, corrosion, or dirt buildup).
- Inspect wiring connections, filters, and components for blockages or leaks.

2 Listen for Noises

- Identify the type of noise (buzzing, rattling, squealing) to narrow down potential faults.
- Isolate whether the sound originates from the indoor or outdoor unit.

3 Measure and Monitor

- Use a thermometer or HVAC tools to check temperature outputs.
- Measure refrigerant pressure and levels to ensure they are within manufacturer specifications.
- Monitor airflow and pressure in the ducts to detect blockages or leaks.

4 Test Components

- Test the thermostat for functionality by changing settings and observing responses.
- Run diagnostic tests on compressors, fans, and motors to assess performance.
- Use multimeters to check electrical continuity and voltage levels.

22.2.5 Conclusion

By recognizing these symptoms and understanding the potential underlying causes, technicians can perform effective troubleshooting, leading to timely repairs and system maintenance.

22.3 MATCHING FAULTS TO SYMPTOMS IN HEATING, VENTILATION, AND AIR CONDITIONING (HVAC) SYSTEMS

Matching faults to symptoms in heating, ventilation, and air conditioning (HVAC) systems is crucial for diagnosing and fixing issues effectively.

22.3.1 Common Faults and Repair Procedures for Heating System and Service Operations

Fault: No Heat or Insufficient Heat

- **Check Thermostat** – Verify the thermostat is set correctly. If it's battery-powered, replace the batteries. Inspect for proper wiring and recalibrate if necessary.
  - ***Service*** – Clean the thermostat sensors, check for loose connections, and ensure the settings match the desired temperature.
- **Check Power Supply** – Inspect the circuit breaker or fuse box for a tripped breaker or blown fuse. Reset or replace if necessary.
  - ***Service*** – Regularly inspect electrical connections and replace any worn-out fuses.
- **Inspect Pilot Light or Ignition** – For gas furnaces, check if the pilot light is out. For electric furnaces, inspect the ignition system.
  - ***Repair*** – Relight the pilot light following safety procedures. If it doesn't stay lit, check the thermocouple and replace if defective. For electric ignitions, inspect and replace faulty components.

Fault: Uneven Heating Across Rooms

- **Inspect Ductwork** – Look for blockages, leaks, or improperly balanced ductwork. Seal any leaks with aluminum tape or mastic sealant.
  - ***Service*** – Clean ducts regularly and adjust dampers for even airflow.
- **Check Air Filters** – Replace clogged or dirty filters that can restrict airflow.
  - ***Service*** – Schedule filter replacements every 3 months or as needed based on usage.

Fault: Short-Cycling (Frequent On/Off)

- **Check Air Filters** – Replace dirty filters that cause overheating and short-cycling.
  - ***Service*** – Keep a regular schedule for filter replacement.
- **Inspect Heat Exchanger** – For possible overheating issues, examine the heat exchanger for cracks or blockages.
  - ***Repair*** – Clean or replace the heat exchanger if it's damaged.

22.3.2 Preventive Maintenance Checklist for Heating Systems

- Replace air filters regularly.
- Clean and inspect ducts for obstructions.

- Lubricate moving parts like blower motors.
- Test thermostat settings for accurate readings.
- Check and clean burners, pilot lights, and ignitions.

22.3.3 Common Faults, Repair Procedures, and Service Operations for Ventilation Systems

Fault: Poor Airflow

- **Inspect Air Filters** – Replace dirty or clogged filters.
  - ***Service*** – Clean or replace air filters every 3-6 months.
- **Inspect Fan and Motor** – Check for wear and tear, lubricate moving parts, and ensure the motor operates smoothly.
  - ***Repair*** – Replace damaged or malfunctioning fan motors.

Fault: Strange Odors Coming from Vents

- **Clean Ducts** – Inspect for mold or mildew and use a vacuum or professional duct-cleaning service to remove contaminants.
  - ***Service*** – Install UV lights or air purifiers to reduce mold growth.
- **Check Drain Pans** – Ensure drain pans are clear of debris and not blocked.
  - ***Repair*** – Clean and sanitize drain pans regularly to prevent odors.

Fault: Inadequate Ventilation or Stuffy Rooms

- **Inspect Vents and Dampers** – Make sure vents are not blocked or dampers are not stuck. Adjust them for proper airflow.
  - ***Service*** – Clean vents regularly and adjust dampers for seasonal changes.

22.3.4 Preventive Maintenance Checklist for Ventilation Systems

- Inspect and replace air filters.
- Clean ducts to remove dust and contaminants.
- Lubricate motor bearings and other moving parts.
- Test dampers for proper functioning.
- Inspect fans and tighten any loose components.

22.3.5 Common Faults and Repair Procedures

Fault: No Cooling or Insufficient Cooling

- **Check Refrigerant Levels** – Use gauges to measure refrigerant pressure. Refill to the recommended level or repair leaks.
  - ***Repair*** – Seal any visible leaks using a sealant or replace damaged pipes. Recharge the refrigerant if needed.
- **Inspect Condenser Coils** – Clean dirty or clogged coils that reduce cooling efficiency.
  - ***Service*** – Use a coil cleaner and a brush to remove debris and improve airflow.
- **Check Compressor** – Inspect the compressor for damage or overheating.
  - ***Repair*** – Replace a faulty compressor if necessary, ensuring proper electrical connections.

Fault: Water Leaking from the Indoor Unit

- **Clear Clogged Condensate Drain** – Use a wet/dry vacuum to clear any clogs in the drain line.
  - ***Repair*** – Replace cracked or damaged drain pans.
- **Check for Frozen Coils** – If evaporator coils are frozen, turn off the AC and let it thaw. Clean dirty coils.
  - ***Service*** – Check refrigerant levels and ensure proper airflow to prevent coil freezing.

Fault: AC Short-Cycling

- **Check Air Filters** – Replace clogged or dirty air filters.
  - ***Service*** – Maintain a regular schedule for filter replacement.
- **Inspect Thermostat** – Ensure the thermostat is functioning properly and recalibrate if needed.
  - ***Repair*** – Replace faulty thermostats if necessary.

22.3.5.1 Table of Fault and Symptom Summary

System	Symptom	Possible Faults
Heating	No heat or insufficient heat	Thermostat issues, blown fuse, clogged filters, pilot light malfunction
Heating	Uneven heating	Unbalanced ductwork, dirty filters, thermostat location

System	Symptom	Possible Faults
Heating	Short-cycling	Oversized unit, dirty filters, thermostat malfunction
Heating	Unusual noises	Loose parts, worn blower motor, debris in ducts
Heating	High energy bills	Leaky ducts, dirty filters, inefficient equipment
Ventilation	Poor airflow	Blocked filters, obstructed ducts, fan motor issues
Ventilation	Strange odors	Mold in ducts, burnt-out components, debris, pest infestation
Ventilation	Excessive dust or allergens	Dirty filters, leaky ducts, inadequate filtration
Ventilation	Inadequate ventilation	Blocked vents, undersized fans, faulty dampers
Air Conditioning	No cooling or insufficient cooling	Low refrigerant, dirty coils, clogged filters, compressor issues
Air Conditioning	Water leaking from indoor unit	Clogged drain line, frozen coils, broken drain pan
Air Conditioning	Unusual noises	Loose parts, fan motor problems, refrigerant leaks, compressor malfunctions
Air Conditioning	Constant running or short-cycling	Oversized unit, blocked airflow, thermostat issues, refrigerant leaks
Air Conditioning	Warm air from vents	Low refrigerant, dirty coils, reversing valve malfunction, fan motor failure

22.3.6 Preventive Maintenance Checklist for Air Conditioning Systems

- Replace air filters every 1-3 months.
- Clean evaporator and condenser coils.
- Inspect and clean the condensate drain line.
- Check refrigerant levels and refill as needed.
- Lubricate moving parts and inspect electrical connections.
- Test thermostat settings and recalibrate for accuracy.

22.3.7 General Tools and Equipment for HVAC Repair and Maintenance

- **Multimeter** – For checking electrical connections.
- **Manifold Gauge Set** – To check refrigerant levels.
- **Vacuum Pump** – For evacuating refrigerant from the system.
- **Fin Comb** – To clean and straighten fins on coils.
- **Pipe Sealant or Tape** – For sealing leaks.
- **Thermometer** – To measure temperature differentials.
- **Screwdrivers and Wrenches** – For adjusting components.
- **Wet/Dry Vacuum** – For cleaning drains and ducts.

22.3.8 Key Safety Precautions for HVAC Repairs

- Always turn off power to the unit before servicing.
- Use personal protective equipment (gloves, goggles) when handling refrigerants.
- Ensure proper ventilation in the workspace when dealing with chemicals.
- Follow manufacturer guidelines for servicing to avoid voiding warranties.
- Be aware of sharp edges on metal ducts and use appropriate tools.

22.3.9 Summary

In summary, understanding and matching these common faults with their symptoms can help in diagnosing issues more accurately, leading to effective repairs and maintenance. Proper maintenance schedules and periodic system checks are also crucial to minimize these faults and ensure optimal performance of HVAC systems.

TO GOD BE

THE GLORY